T0292078

This invaluable reference handbook describes the fundamental principles and procedures underlying the successful isolation of viable, functionally intact haematopoietic and lymphoid cells, and their maintenance as primary cultures. It provides the technical information on the signals and mediators required for the differentiation and growth of these cells, and is designed for laboratory investigators with limited practical experience in cell culture. There are chapters on dendritic cells, T and B lymphocytes, monocytes and macrophages, NK and LAK cells, mast cells and basophils, as well as on haematopoietic differentiation of embryonal stem cells and on culturing murine thymic explants. Each chapter had been written by experts who have direct practical experience of the techniques and can therefore provide tips for the avoidance of common pitfalls, as well as an insight into the fundamental questions in cell biology and immunology which can be addressed using each cell culture model.

MAGGIE DALLMAN has worked in the application of basic immunological research to the area of organ transplantation for over two decades. Her interests have included the cellular components of graft rejection, cytokines in transplantation and molecular interactions controlling the immune response. She is currently Professor of Immunology at Imperial College of Science, Technology and Medicine, London.

JONATHAN LAMB holds the Glaxo-Wellcome/British Lung Foundation Chair of Respiratory Science at the University of Edinburgh. His scientific research has focused on the biology of T lymphocytes with particular interest in immune regulation and immunological tolerance and their application in chronic pulmonary inflammation.

Handbooks in Practical Animal Cell Biology

Series editor:
Dr Ann Harris
Institute of Molecular Medicine, University of Oxford

The ability to grow cells in culture is an important and recognised part of biomedical research, but getting the culturing conditions correct for any particular type of cell is not always easy. These books aim to overcome this problem. The conditions necessary to culture different types of cell are clearly and simply explained in seven different volumes. Each volume covers a particular type of cell, and contains chapters by recognised experts explaining how to culture different lineages of the cell type. There is also a volume on general techniques in cell culturing. These practical handbooks are clearly essential reading for anyone who uses cell culture in the course of their research.

Also in the series:

General techniques of cell culture, by M. Harrison and I. Rae

Epithelial cell culture, edited by A. Harris

Endothelial cell culture, edited by R. Bicknell

Marrow stromal cell culture, edited by J. N. Beresford and M. E. Owen

Endocrine cell culture, edited by S. Bidey

Haematopoietic and lymphoid cell culture

Edited by

Margaret J. Dallman

and

Jonathan R. Lamb

CAMBRIDGE
UNIVERSITY PRESS

CAMBRIDGE UNIVERSITY PRESS
Cambridge, New York, Melbourne, Madrid, Cape Town, Singapore, São Paulo

Cambridge University Press
The Edinburgh Building, Cambridge CB2 2RU, UK

Published in the United States of America by Cambridge University Press, New York

www.cambridge.org
Information on this title: www.cambridge.org/9780521620437

© Cambridge University Press 2000

First published 2000

A catalogue record for this publication is available from the British Library

Library of Congress Cataloguing in Publication data

Haematopoietic and lymphoid cell culture / editors, Maggie Dallman and
  Jonathan Lamb.
      p.   cm. – (Handbooks in practical animal cell biology)
    ISBN 0 521 62043 0. – ISBN 0 521 62969 1 (pbk.)
      1. Hematopoietic stem cells Handbooks, manuals, etc.
    2. Immunocompetent cells Handbooks, manuals, etc.   3. Cell culture
    Handbooks, manuals, etc.   I. Dallman, Maggie, 1957–   .   II. Lamb,
    Jonathan R.   III. Series.
    QP92.H34   2000
    612.4′1–dc21   99-38930 CIP

ISBN-13 978-0-521-62043-7 hardback
ISBN-10 0-521-62043-0 hardback

ISBN-13 978-0-521-62969-0 paperback
ISBN-10 0-521-62969-1 paperback

Transferred to digital printing 2006

Contents

Contributors

Margaret J. Dallman (Editor)
Department of Biology, Imperial College of Science, Technology and Medicine, Sir Alexander Fleming Building, Imperial College Road, London SW7 2AZ, UK

Jonathan R. Lamb (Editor)
Respiratory Medicine Unit, University of Edinburgh Medical School, Teviot Place, Edinburgh EH8 9AG, UK

Kim L. Anderson
Anatomy Department, Medical School, University of Birmingham, Birmingham B15 2TT, UK

John Ansell
John Hughes Bennet Laboratory, Department of Medicine, Western General Hospital, Crewe Road, Edinburgh EH4 2XU, UK

Jonathan M. Austyn
Nuffield Department of Surgery, University of Oxford, John Radcliffe Hospital, Headington, Oxford OX3 9DU, UK

Colin G. Brooks
School of Microbiological, Immunological and Virological Sciences, The Medical School, University of Newcastle upon Tyne, Framlington Place, Newcastle upon Tyne NE2 4HH, UK

David Chao
Nuffield Department of Surgery, University of Oxford, John Radcliffe Hospital, Headington, Oxford OX3 9DU, UK

Anne E. Corcoran

Medical Research Council, Laboratory of Molecular Biology, Hills Road, Cambridge CB2 2QH, UK

Current address: Cambridge Institute of Medical Research, Department of Oncology, Level 6, Wellcome Trust/MRC Building, Addenbrookes Hospital, Hills Road, Cambridge CB2 2XY, UK

Siamon Gordon

Sir William Dunn School of Pathology, University of Oxford, South Parks Road, Oxford OX1 3RE, UK

Richard Haworth

Sir William Dunn School of Pathology, University of Oxford, South Parks Road, Oxford OX1 3RE, UK

Nicholas Hole

Department of Biological Sciences, University of Durham, South Road, Durham DH1 3LE, UK

Chen-Lung Lin

Nuffield Department of Surgery, University of Oxford, John Radcliffe Hospital, Headington, Oxford OX3 9DU, UK

James A. Mahoney

Sir William Dunn School of Pathology, University of Oxford, South Parks Road, Oxford OX1 3RE, UK

Dean D. Metcalfe

Laboratory of Allergic Diseases, NIAID, National Institutes of Health, Bethesda MD 20892-001, USA

Gunnar Nilsson

Department of Genetics and Pathology, Uppsala University, S-751 85 Uppsala, Sweden

John J. T. Owen

Anatomy Department, Medical School, University of Birmingham, Birmingham B15 2TT, UK

Justin A. Roake

University Department of Surgery, Christchurch School of Medicine, Christchurch Hospital, New Zealand

Hergen Spits

The Netherlands Cancer Institute, Division of Immunology, Plesmanlaan 121, 1066 CX, Amsterdam, The Netherlands

Rakesh Suri

Nuffield Department of Surgery, University of Oxford, John Radcliffe Hospital, Headington, Oxford OX3 9DU, UK

Ashok R. Venkitaraman

Medical Research Council, Laboratory of Molecular Biology, Hills Road, Cambridge CB2 2QH, UK

Current address: Cambridge Institute of Medical Research, Department of Oncology, Level 6, Wellcome Trust/MRC Building, Addenbrookes Hospital, Hills Road, Cambridge CB2 2XY, UK

Preface to the series

The series Handbooks in Practical Animal Cell Biology was born out of a wish to provide scientists turning to cell biology, to answer specific biological questions, the same scope as those turning to molecular biology as a tool. Look on any molecular cell biology laboratory's bookshelf and you will find one or more multivolume works that provide excellent recipe books for molecular techniques. Practical cell biology normally has a much lower profile, usually with a few more theoretical books on the cell types being studied in that laboratory.

The aim of this series, then, is to provide a multivolume, recipe-book-style approach to cell biology. Individuals may wish to acquire one or more volumes for personal use. Laboratories are likely to find the whole series a valuable addition to the 'in house' technique base.

There is no doubt that a competent molecular cell biologist will need 'green fingers' and patience to succeed in the culture of many primary cell types. However, with our increasing knowledge of the molecular explanation for many complex biological processes, the need to study differentiated cell lineages *in vitro* becomes ever more fundamental to many research programmes. Many of the more tedious elements in cell biology have become less onerous due to the commercial availability of most reagents. Further, the element of 'witchcraft' involved in success in culturing particular primary cells has diminished as more individuals are successful. The chapters in each volume of the series are written by experts in the culture of each cell type. The specific aim of the series is to share that technical expertise with others. We, the editors and authors, wish you every success in achieving this.

ANN HARRIS

Introduction

Cells of the haematopoietic system are critical in defense of the body against invading microorganisms. They form a complex system of interacting cell types that mediate innate and acquired immunity and inflammatory reactions. An ability to grow many of these cell types in their immature and/or mature states has contributed greatly not only to our understanding of the development of individual cell types, but also of their interactions in the animal.

In this volume characteristics of the major haematopoietic cell types are given together with their isolation and growth requirements, in most cases for both progenitors and mature cells. The great diversity of functional and phenotypic characteristics of cells of the haematopoietic system is reflected in their very different growth and maintenance requirements. The isolation of multi-potent haematopoietic stem cells is not covered in this volume, but the early differentiation of haematopoietic cells from embryonic stem cells is included.

1. *Ansell and Hole*: Haematopoietic differentiation of embryonic stem (ES) cells

Embryonic stem cells have provided a critical resource in the development of mice carrying defined mutations in their genome. However, in an as yet less widely used application, they are proving to be an invaluable tool in the dissection of the very earliest stages of cell differentiation. In Chapter 1 the potential of such cells in the analysis of early differentiation of haematopoietic stem cells is described. Optimal conditions for the growth of ES cells in an undifferentiated state for such studies are discussed together with details of methods for their aggregation and differentiation. Methods for analysis of differentiated progeny using both phenotype and functional approaches are included.

2. Austyn, Chao, Lin, Roake and Suri: Dendritic cells (DCs)

The importance of the dendritic cell in the initiation of T cell dependent immune responses is now well established. This has led to enormous interest in their use in vaccination protocols both for experimental and clinical purposes. More recently it has become clear that DCs may also have a role in regulation of immunity by interacting with T cells in different ways or at different stages of their development. Until relatively recently full investigation of both the phenotypic and functional characteristics of DCs has been hampered by their relative paucity in all tissues, meaning that their isolation has been particularly painstaking. Lack of truly specific cell surface markers has only added to these problems. An ability to grow such cells from bone marrow, blood and other tissues, as described in Chapter 2, has indicated that culture conditions determine both phenotype and function of the resulting populations. This chapter discusses requirements and characteristics of both mouse and human DCs together with techniques that may be used in the study of their migration characteristics.

3. Anderson and Owen: Murine thymic explant cultures

The differentiation of thymocytes into mature antigen–specific T lymphocytes has been the focus of intense investigation over many years. Whilst the use of bone marrow chimeras and more recently transgenic/knockout animals has contributed enormously to our understanding of many of the events occurring during maturation and selection, it has been difficult to analyse certain molecular events and thymocyte–stromal cell interactions in whole animals. The development of thymic organ and re–aggregation cultures as described in Chapter 3 has allowed such studies to be performed under closely controlled conditions thus vastly increasing knowledge of such events.

4. Spits and Brooks: T cells

T cells play a central role in the development of acquired immunity. Following an interaction with their specific MHC–peptide complex they become activated, differentiating further into effector cells and providing growth and differentiation signals for many other cells of the haematopoietic system. The functional, structural and genetic analysis of T cells representative of different subsets has been possible only through advances in our knowledge of their long-term growth requirements. This has allowed stable cultures of T cells to be established. In Chapter 4, Spits and Brooks discuss technical aspects underlying the isolation and expansion of T cells lines and clones. The information they provide will be valuable in addressing a wide range of questions on T cell biology and the study of human disease.

5. *Corcoran and Venkitaraman*: B lymphocytes

B lymphocytes are critical in the adaptive immune response not only for their ability to differentiate into the antibody-producing plasma cell, but also for their interactions with T lymphocytes. The development of B cell memory, an as yet incompletely defined process, forms the basis of many vaccination protocols. Chapter 5 provides an accessible and comprehensive introduction to our understanding of B cell development. Conditions of culture required for early bone marrow derived B cells through to B cells of the spleen and peritoneal cavity in both human and mouse are covered together with techniques for their phenotypic analysis using both flow cytometry and molecular techniques.

6. *Mahoney, Haworth and Gordon*: Monocytes and macrophages

Phagocytes of the innate immune system including macrophages provide a first-line defence against invading microorganisms. They bear surface receptors, such as the mannose receptor, which recognise bacteria triggering phagocytosis and the secretion of soluble mediators. Such mediators are involved in both inflammation and generation of specific immune responses. In Chapter 6, methods for the isolation and culture of primary monocytes and macrophages from a variety of tissues in both mouse and human are described. These techniques will allow the full functional repertoire of these cell types in both innate and acquired immunity to be investigated. Previous analysis had been restricted to cell lines displaying restricted characteristics of the monocytic lineage.

7. *Brooks and Spits*: Natural killer (NK) cells and lymphokine activated killer (LAK) cells

NK cells, despite being well represented in the haematopoietic system and having been discovered about 20 years ago, are relatively poorly characterised with respect to their physiological role. Recently, however, much progress has been made in our understanding of their surface recognition structures. It has become clear that MHC recognition is associated with negative signalling and it is the absence of MHC which triggers a lytic response.

Chapter 7, on NK and LAK cells, begins with the definition and characterisation of these cells in relation to conventional T cells. The authors then
provide a detailed description of the growth requirements of NK cells high-
lighting the difficulty in cloning and expanding monoclonal populations of
NK cells from adult mice.

8. *Nilsson and Metcalfe*: Mast cells and basophils

Mast cells and basophils are effector cells that originally evolved to provide
protection against intestinal worms. However, their ability to bind IgE, an
effector molecule induced following exposure to allergen in genetically susceptible individuals, generates a pathogenic response. Through the cross-
linking of their membrane-bound IgE receptors by specific antigen, these
cells release an array of preformed and newly synthesised mediators.
Investigators wishing to explore different biological aspects of the principal
effector cells in immediate hypersensitivity reactions will find Chapter 8
describing the culture conditions for mast cells and basophils very useful.
The identification of IL–3 and stem cell factor as growth factors for these cell
types has made *in vitro* culture possible.

1

Haematopoietic differentiation of embryonic stem cells

John Ansell and Nicholas Hole

Introduction

All of the blood cell lineages differentiate from a pool of haematopoietic stem cells (HSC). These cells, although in a very small minority in the bone marrow of adult mammals, are highly pluripotent, not only capable of differentiating into all of the mature cell types (platelets, mast cells, neutrophils, mono-cyte/macrophages, eosinophils, erythrocytes, thymocytes and B cells) of the haematopoietic system but also of 'self-renewing', maintaining and if necessary expanding their numbers throughout life. HSC also provide the long-term repopulating ability of bone marrow after its transplantation, say, into an irradiated or genetically compromised host. Thus bone marrow transplantation has been the preferred and sometimes the only option for patients undergoing intensive radio- and/or chemotherapy not only for aggressive leukaemia and related cancers but also in the treatment of blood disorders. However, the extreme difficulties in tracing matched donors in sufficient numbers have led to a search for alternative sources of HSC. In the last five years, two alternatives have been developed: peripheral blood stem cells and umbilical cord blood stem cells. In the former the very small numbers of HSC found in adults' peripheral blood can be increased by 'mobilisation' treatments and collected from a blood donation. In the latter there are proportionately higher numbers of stem cells in fetal blood, which can be collected from the umbilical cord at delivery. In both cases the stem cells have to be enriched from the blood and their numbers expanded before transplantation. There is increasing evidence that the expansion of stem cell numbers using growth factors and cytokines reduces their self-renewal properties and thus their potential for long-term reconstitution of the blood (Holyoake *et al.*, 1997).

These developments have great potential to provide a much more readily available source of transplantable HSC and there is a need to exploit their development rapidly.

However, studies of the physiological and molecular/genetic control of HSC proliferation and differentiation have been hampered by several factors:

1 Their paucity of numbers *in vivo* is compounded by an imprecise definition of their phenotype, making specific isolation of HSC difficult.
2 Derivation of haematopoietic cell lines *in vitro* is dependent upon (usually retroviral) transformation. Those lines that have been developed have limited pluripotency and their maintenance is usually only possible in the presence of cytokines and/or growth factors.
3 Genetic manipulation of HSC is possible but this is typically reliant on retroviral infection.

The lack of detailed knowledge of the mechanisms controlling stem cell differentiation and self-renewal profoundly affects our ability to expand HSC numbers without their concomitant differentiation and reduction in pluripotentiality although clearly this HSE expansion process must occur at some point during embryonic development.

One class of stem cells that can be maintained and expanded *in vitro* without losing *in vivo* pluripotency is murine embryonic stem (ES) cells. These cells are derived from the inner cell mass of early (3.5 day) mouse blastocysts and can be maintained indefinitely as undifferentiated cell lines *in vitro* in the presence of a polypeptide cytokine – leukemia inhibitory factor (LIF) (Smith *et al.*, 1988). After their introduction back into pre-implantation embryos, ES cells have the capacity to contribute to all tissue systems and, crucially, to the germ line of the resulting animal. It has been formally demonstrated that ES cells can generate pluripotent HSC capable of long-term reconstitution (Forrester *et al.*, 1991). This was achieved using techniques of aggregation of ES cells with tetraploid host embryos. Mice derived by this process are entirely from the ES cell contribution to the original aggregation, tetraploid cells only having contributed to extra-embryonic structures. Although peri-natal death is the most likely outcome of this procedure, Forrester *et al.* (1991) demonstrated that the fetal livers of such mice contained HSC capable of entirely repopulating the haematopoietic system of irradiated secondary recipients. All of the cell types of the blood of the reconstituted host were shown to have been ES-derived.

Importantly for haematologists ES cells also have the capacity to differentiate *in vitro* into a variety of different cell types, including those of the haematopoietic system, albeit in a relatively disorganised way. These properties of ES cells have excited the interest of stem cell biologists, not only because they facilitate the opportunity to study the 'rules' of stem cell growth and self-maintenance, but also because of the ease with which ES cells can be genetically manipulated prior to their re-introduction *in vivo*. Through this process it becomes possible to generate a range of animal models of human disease and malignancy.

Thus the ability to control and direct the differentiation of ES cells into haematopoietic lineages could have several advantages:

1 As a potential source of long-term haematopoietic repopulating cells to be used for transplantation into compromised hosts.
2 As an *in vitro* test system for protocols designed to expand stem cell numbers.
3 As a model system in which the effects of transgenic modifications could be screened prior to their introduction into the germ-line.
4 As a system in which the molecular events controlling the differentiation and maintenance of HSC could be delineated.

As mentioned above, when LIF is removed from the culture medium ES cells will differentiate, giving rise to various endodermal, mesenchymal and neuronal cells. As part of this process the differentiation of haematopoietic cells is now well established and erythrocytes, macrophages, lymphocytes, neutrophils, and mast cells have all been observed during ES cell differentiation *in vitro* (Doetschman *et al.*, 1985; Wiles & Keller, 1991; Chen, Kosco & Staerz, 1992). Initially only long-term lymphopoiesis in mice in receipt of differentiated ES cells had been achieved (Chen *et al.*, 1992) suggesting that repopulating potential after ES cell differentiation was limited to the production of lymphoid progenitors. Subsequently we and others have demonstrated that multilineage repopulation of mice using ES-derived progenitors is possible (Palacios, Golunski & Samardis, 1995; Hole *et al.*, 1996*a*). The production *in vivo* of differentiated haematopoietic cells derived from pluripotential ES cells implies that *in vitro* self-renewing haematopoietic precursors must be present at some point following the withdrawal of LIF. The protocols for the differentiation and identification of such haematopoietic progenitors from ES cells are the subjects of this chapter.

General notes on methodologies and reagents

In all of the methodologies the water used is either 'Analar' grade (BDH) or Milli-Q (Millipore) purified.

ES cell culture and differentiation medium: the only difference between these two media is the batch of serum used and the absence of LIF in the latter. We are currently using fetal calf serum from Globepharm in the ES cell culture medium and from Boehringer in the differentiation medium. Regardless of the source of serum, the potential supplies have to be batch-tested to check for the least differentiation effects in the case of ES cell culture and, for example, for maximum CFU-A yielding potential in differentiation cultures. Similarly the donor horse serum used in the CFU-A assay should be batch-tested before use (see Pragnell, Freshney & Wright, 1994)

Reagents

10× Glasgow's modified Eagles medium (GMEM)★
10× α-Modified Eagles medium (αMEM)★
Horse serum★
Sodium bicarbonate★
Non-essential amino acids★
Glutamine/pyruvate★
Chicken serum★
0.1 M 2-mercaptoethanol (BDH)
PBS tablets (Oxoid)
Collagenase (Difco)
Gelatin (porcine skin gelatin Type A) (Sigma)
Noble agar (ICN Flow)
Trypsin (ICN Flow)
EDTA disodium salt (Fischer Scientific UK)
Methyl cellulose; 15 centipoises viscosity (Metachem Diagnostics Ltd)

★ From GIBCO–BRL

Embryonic stem cell isolation and culture

The derivation of ES cell lines and their routine culture has been described in extensive detail by Smith (1991) and will not be specifically dealt with in any detail in this chapter. These methodologies describe ES cell culture in the absence of feeder cells using LIF supplemented medium. Although some laboratories continue to grow ES cells on feeder layers these protocols are generally inconvenient for ES cell differentiation. Inadequate mitotic inactivation and the possibility of feeder cell inclusion into differentiating ES cell structures are particular problems.

ES cell culture

The ES cell lines that have been used in the protocols described here include EFC-1 (Nichols, Evans & Smith, 1990) and CGR8 (ECACC 95011018) although the methods can equally be applied to most ES cell lines. In brief, ES cells are maintained in GMEM supplemented with 10% fetal calf serum, 1% non-essential amino acids, 2% L-glutamine/sodium pyruvate (from a stock solution containing a 1:1 volume ratio of 200 mM glutamine and 100 mM pyruvate), 1×10^{-4} M 2-mercaptoethanol and 0.2% sodium bicarbonate. To maintain ES cells in their undifferentiated state 100 U/ml of LIF (obtained as culture supernatant from Cos-7 cells transfected with a LIF gene-containing expression vector (Smith, 1991)) is added to the culture medium. 1 unit (U) of LIF is defined as that volume of transfected Cos-7 supernatant in 1 ml of medium giving rise to detectable inhibition of ES cell differentiation. Cells are incubated at 37 °C in a humidified atmosphere of 5% CO_2 in air usually in 25 cm² flasks coated with 1% gelatin.

ES cell harvesting

ES cells are harvested when sub-confluent.

1 After aspiration of the spent culture medium the cells are washed gently with 5 ml PBS.
2 Cells are released into a single cell suspension by the addition of 1 ml of TVP (1% chick serum, 1% trypsin, 8.4×10^{-4} M EDTA) and incubation at 37 °C for 2 min.
3 The cell suspension is harvested after physical agitation, 9 ml of culture medium added and the suspension spun at 100g for 5 min.
4 After aspiration of the supernatant and resuspension in 10 ml culture medium the cells are counted in a haemocytometer.
5 A similar procedure is used should cells need to be frozen at this stage except that cells are harvested in 7 ml of medium spun and resuspended in 1 ml of 10% DMSO (v/v) in ES cell culture medium. 0.5 ml of suspension is then pipetted into a cryo-tube and frozen at -70 °C overnight prior to storage in liquid nitrogen.

From a sub-confluent 25 cm² flask, one should typically expect a yield of ~ 5–9×10^6 cells.

Generation of undifferentiated ES cell aggregates

Studies of the differentiation of mesodermal lineages from ES cells have usually relied on the formation of aggregates of ES cells (see Wiles, 1993; Hole & Smith, 1994) prior to the formation of embryoid bodies (see below). Such aggregates have been formed by enzymatic digestion of monolayer cultures (Schmidt, Bruyns & Snodgrass, 1991), direct formation in semi-solid medium (Keller et al., 1993; Wiles, 1993) and in hanging drops (Hole & Smith, 1994). This last technique is the one currently in use in our laboratories for the differentiation of transplantable haematopoietic progenitors and will be described in some detail. After the removal of LIF, aggregates will differentiate in situ in methyl cellulose, in suspension culture or on stromal layers. A proportion of these aggregates form 'embryoid bodies' (EBs), complex structures with large fluid-filled cysts that have a superficial similarity to embryonic yolk sacs. These structures contain cells of a number of different lineages, such as muscle, endothelium and cells of the haematopoietic system: evidence for haematopoietic cells is often indicated by the presence of haemoglobinised blood islands.

The benefits of one differentiation system over another are a matter of some debate. Haematopoietic differentiation in suspension culture or in semi-solid medium may well be dependent on the formation of EBs (Doetschman et al., 1985) and while some workers have claimed that formation of cystic structures may be a pre-requisite for complete differentiation (Chen et al., 1992; Hooper, 1994), this may not be an absolute requirement in the case of haematopoietic differentiation (Doetschman et al., 1985; Schmidt et al., 1991). Differentiation in EBs can be followed for extended periods (>30 days), but two to three weeks is a more normal time period.

Hanging drop method

Although this method may be time-consuming (Fig. 1.1) it produces, prior to differentiation, ES cell aggregates of a uniform size. This may be important for subsequent differentiation into embryoid bodies since endogenously produced LIF may exert some residual autocrine or paracrine function for a time after removal of exogenous LIF.

1 Adjust the density of the cell harvest (see above) to $\sim 3 \times 10^4$ cells/ml with culture medium and add 100 U/ml of LIF.
2 Add 10 ml water to the base of a (square) bacteriological plate. Turn the lid upside down and using an Eppendorf multipipette, or its equivalent, dispense 10 µl drops of cell suspension (approx. 300 cells) on the now upturned underside of the lid taking care to keep sufficient distance between drops to avoid coalescence; approximately 100 drops can be placed on one lid.
3 With a smooth, swift action, replace the lid on its base and carefully place

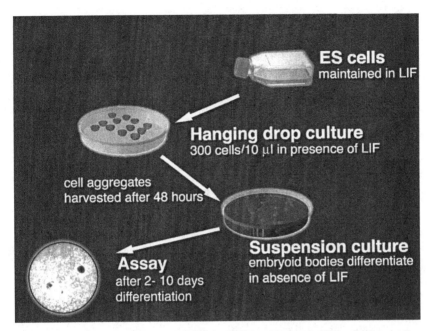

Fig. 1.1. Schematic diagram of haematopoietic differentiation of embryonal stem cells *in vitro*: hanging drop method.

the dish in a humidified incubator in an atmosphere of 7.5% CO_2 in air. Culture the hanging drops for 48 h.

4 After two days the ES cells will have settled in the meniscus of the hanging drop forming smooth, spherical aggregates of a uniform size.

5 With cell division these aggregates have relatively little structure and are sensitive to mechanical disruption but will now contain ~1500 cells, the majority of which should be in an undifferentiated state. They are therefore harvested with care by aspiration with a wide bore pipette and placed in a 15 ml centrifuge tube and centrifuged at 80g for 3 min.

6 Aspirate the supernatant from the aggregates and wash twice in 5 ml differentiation medium.

High density culture

A more rapid method for generating large numbers of aggregates is to culture ES cells at high density in non-gelatinised flasks. The ES cells form aggregates that are only loosely attached to the substrate and these can be harvested directly. Aggregates prepared in this way may vary in size and their subsequent differentiation is also variable.

1 Seed a 25 cm^2 non-gelatinised tissue culture flask with 2×10^6 ES cells in 10 ml of pre-warmed culture medium in the presence of LIF.
2 Culture for two days in humidified 7.5% CO_2 incubator.
3 Depending on the growth rate of the ES cell line used, the medium may have to be changed after one day. If so, simply carefully replace spent medium with fresh culture medium and LIF. In replacing the medium, be careful not to disturb the attached cells at this point.
4 The aggregates are then harvested directly in a similar way to that described above.

Formation of embryoid bodies (EBs)

Differentiation in suspension culture

1 Re-suspend ES cell aggregates formed as described above at a concentration of ~100 aggregates/ml in differentiation medium containing 50 units/ml of penicillin and 50 µg/ml of streptomycin.
2 The suspension is cultured in 9 cm microbial Petri dishes in a humidified atmosphere of 7.5% CO_2 in air. The medium should be changed every two days by allowing EBs to settle and aspirating supernatant.
3 The aggregates do not attach to the Petri dishes, but as the EBs begin to differentiate they adhere to the substrate, and can be detached by gentle aspiration. The EBs that develop can form very large cystic structures (up to ~6 mm in size). (See Fig. 1.2).

Differentiation in semi-solid medium

Suspension culture of EBs is appropriate where the haematopoietic cells under study are immobile or only poorly motile (such as HSC or red blood cells [RBC]). However, where motile haematopoietic cells are being investigated, identifying which EBs are producing (for example) macrophages may require a method that restricts dispersion of EBs and their progeny. One approach is to culture aggregates in semi-solid medium, although the difficulties of changing the medium in this system militates against extended (>15 days) culture. The semi-solid medium of choice in our laboratory contains 0.3% Noble agar.

1 Prepare 0.3% Noble agar by mixing equal volumes of freshly prepared 0.6% (2×) agar stock (2.4 g of agar boiled in 200 ml of tissue culture grade water) which has been cooled to 56 °C in a water bath and 2× differentiation medium pre-warmed to 37 °C.

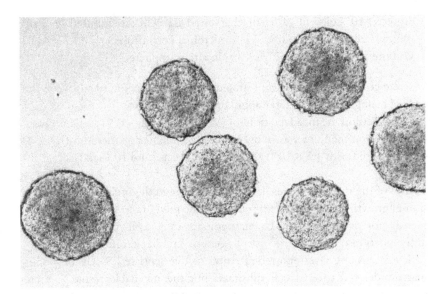

Fig. 1.2. Embryoid bodies after 2 days of culture in the absence of LIF.

2 Stir the mixture and allow to equilibrate at 37°C.
3 Approx. 100–500 ES cell aggregates should be re-suspended in 10 ml agar medium, transferred to a 9 cm microbial Petri dish and cultured as above.

If necessary it is possible to 'feed' the EBs by gently overlaying the culture with 5 ml fresh differentiation medium.

Direct differentiation in methyl cellulose

Direct plating of ES cells into a medium containing methyl cellulose allows ES cells to form EBs and undergo haematopoietic differentiation without the prior formation of ES cell aggregates. This method is quicker than those described above and it may be more appropriate for other ES cell isolates, but in our hands it does not improve differentiation and is still subject to the problems of culture in semi-solid medium.

1 Methyl cellulose medium is prepared by mixing an equal volume of 2× methyl cellulose stock (3.6 g of methyl cellulose dissolved in 200 ml of gently boiling tissue culture grade water and then allowed to to cool in a 37°C water bath) with pre-warmed 2× differentiation medium.
2 The medium is stirred and allowed to equilibrate.

3 Approx. 10^3 cells/ml of freshly harvested ES cells are cultured in 2 ml methyl cellulose medium in 4 cm microbial Petri dishes.

4 Culture in a humidified 7.5% CO_2 in air atmosphere.

Discrete colonies of ES cells and subsequent EBs will begin to appear in the methyl cellulose medium after approximately three days.

This method is an adaptation of Wiles & Keller (1991). Their basic differentiation medium was Iscove's modified Dulbecco's medium (IDDM) supplemented with 4.5×10^{-4} M monothioglycerol and 10% fetal calf serum (FCS).

Any of the methodologies described above lend themselves to the inclusion of growth factors and/or cytokines in the medium to study the requirements, for example, for haematopoietic stem cell proliferation and differentiation or to alter the morphogenetic microenvironment generally.

Several groups have reported attempts to differentiate ES cells on bone-marrow-derived stromal cell substrates but the inevitable result of these experiments was preferential differentiation into cells of macrophage and monocyte lineages. This was thought to be due to the secretion of macrophage colony stimulating factor (M-CSF). More recently a Japanese group (Nakano, Kodama & Honjo, 1994) have used the OP9 stromal cell line, which was developed from an osteopetrotic mouse mutant [(C57BL/6×C3H)F2-op/op], which constituitively lacks functional M-CSF secretion. In these experiments ES cells were seeded directly onto the stromal layer in suspension culture. After passage the ES cells were shown to have differentiated into a variety of cell types of the lymphohaematopoietic system with the formation of multilineage colonies. This system was characterised by the absence of embryoid body-like structures.

Analysis of differentiated progeny

Obviously once haematopoietic differentiation has been achieved by any of the above techniques many forms of analysis are possible. These include histological examination, flow cytometric analysis, gene expression studies, analyses of haematopoietic colony-forming potential and studies of the ability of the differentiated progeny to reconstitute the haematopoietic system of, say, irradiated recipient animals.

Histology

Simple histological analysis of cystic EBs has been used to define the appearance of blood islands, neuroepithelium, mesoderm and visceral and parietal endoderm (Doetschman *et al.*, 1985; Hooper, 1994) and some groups have claimed that distinct areas of the embryoid bodies allow development of lymphoid cells in particular (Chen *et al.*, 1992). Erythroid cells were the first haematopoietic cell type to be described in differentiating embryoid bodies (Doetschman *et al.*, 1985) using light microscopy to observe haemoglobinisation directly. The red colour associated with erythropoiesis appears in EBs from 5 to 14 days after the initiation of differentiation. Quantitation of this phenomenon is a non-destructive and rapid means of determining the effects of culture conditions on haematopoietic development but does not in itself convey much information about other haematopoietic lineages.

Flow cytometric analysis

Flow cytometric analyses can be used to follow the differentiation of haematopoietic precursors and their progeny during EB differentiation. Although in mice a precise cell surface phenotype of HSCs has yet to be defined, most of the later lineages can be distinguished by the expression of specific antigens. The presence of B cells (B220 + ve), erythroid lineage cells (TER 119 + ve) and myeloid (Mac-1 + ve) cells have all been detected by flow cytometry, although in some laboratories only small numbers of cells expressing the haematopoietic lineage marker CD45 have been seen and those cells expressing neutrophil, macrophage and B cell markers have not been detected (Schmidt *et al.*, 1991). Problems arising with flow cytometric analyses include:

1 Embryoid bodies contain very few haematopoietic cells at least at early stages of differentiation and these may often be at the limits of flow cytometric detection.
2 The antigens that define lineages in the context of normal haematopoiesis may be expressed by non-haematopoietic lineages within the embryoid bodies. For example, ES cells possess low-level mRNA transcripts for Thy-1, CD44 and Ly-6 (Schmidt *et al.*, 1991; Hole *et al.*, 1996*a*,*b*). The appearance of these markers on the surface of differentiated progeny may thus not define them as haematopoietic.
3 During the preparation of single-cell suspensions from EBs, several groups commonly use enzymatic digestion protocols. This kind of treatment may alter the cell surface expression of markers by enzymatic degradation.

We have used two methods to derive single-cell suspensions from EBs to circumvent problem (3):

Homogenisation

1 EBs are harvested and washed twice in PBS by allowing the bodies to settle and aspirating the supernatant.
2 They are then resuspended in 1–2 ml of 1 mM EDTA in PBS and left at room temperature for 5 min.
3 After transfer to a glass or Teflon homogeniser, 5 strokes of the plunger will disrupt the EBs.
4 The supension is transferred to a 15 ml centrifuge tube and left to allow cellular clumps to settle out.
5 The single-cell suspension in the supernatant is harvested and washed in 10 ml PBS containing 1% heat-inactivated fetal calf serum by centrifugation at 100g for 5 min.
6 After re-suspension the cell count and viability is checked by trypan blue dye exclusion. The cell yield should be 1–5 × 10^3 cells/embryoid body, depending on time of differentiation, with a viability of 70–80%.

Enzyme digestion

1 Embryoid bodies are harvested and washed twice in PBS by allowing the bodies to settle and aspirating the supernatant.
2 Resuspend in 2 ml of TVP, transfer to a 15 ml centrifuge tube and incubate for 20 min at 37°C on a roller.
3 Break up the remaining EB structures by gently pipetting through a Pasteur pipette.
4 Leave the suspension to settle for 5 min.
5 The supernatant containing a single-cell suspension is then harvested leaving behind any remaining clumps and non-cellular proteinaceous material, transferred to 10 ml of culture medium and centrifuged at 100g for 5 min.
6 Repeat the wash step in PBS containing 1% heat-inactivated PBS.
7 Cell counts and viability are checked by trypan blue exclusion.

Viability using this procedure should be between 80 and 90%. An alternative to TVP is to use Dispase (Boehringer Mannheim). EBs are incubated in PBS containing 1 unit/ml of Dispase for one hour with gentle agitation, aspirated several times through a 23-gauge needle and then washed as before. This has been our method of choice for repopulation studies (see below).

Gene expression studies

An alternative to flow cytometric analyses for many laboratories has been to follow the patterns of temporal and/or lineage restricted expression of mRNA transcripts in differentiating EBs using reverse transcriptase–polymerase chain reaction (RT-PCR) and Northern blotting. A detailed description of the methodologies involved and the data obtained through this approach is beyond the scope of this chapter and has been reviewed elsewhere (Hole & Ansell, 1996). Although this kind of approach does allow studies of developmental regulation of EB differentiation it is invasive and is dependent on the preparation of mRNA from bulk cultures. In the future, *in situ* expression analysis in individual EBs and visualisation with, say, confocal microscopy, may allow studies of the simultaneous expression of genes involved in haematopoietic differentiation both temporally and spatially. Thus it may be possible to relate gene expression in haematopoietic cells within EBs with co-ordinate expression in stromal elements in the vicinity. Attempts are being made to combine flow cytometric analyses with gene expression studies (Ansell & Hole, unpublished) by using ES cells made transgenic for fluorescent intracellular markers such as β-galactosidase or Green Fluorescent Protein, driven by the controlling elements of genes known to be expressed early during haematopoietic differentiation. These cells can then be sorted from bulk cultures and subjected to further molecular/biochemical analyses obviating the problems (described above) inherent in conventional flow cytometric analyses.

Analysis of colony forming potential

Characterisation of haematopoietic progenitors has been possible by the use of *in vitro* colony forming assays. Several of these have been used to confirm that the programme of ES cell differentiation follows the normal route of haematopoietic development and the expected spectrum of progenitors and cells from the multipotential to the terminally differentiated has been detected. Mixed-cell, erythroid and granulocyte/macrophage colony forming units have all been detected, first appearing at or around day 4 of differentiation, before the appearance of markers of their respective terminally differentiated progeny and thereafter decreasing in number as differentiation procedes. These data indicate that the appearance of the most primitive haematopoietic progenitors may be an early and transient event during embryoid body differentiation. Although there are no colony forming unit assays that will detect the highly pluripotent HSCs that will effect long-term repopulation, there are several that can be used to detect multipotent progenitors. We have used one of these, the CFU-A assay (Pragnell *et al.*, 1994), which detects a primitive progenitor that has similar cycling characteristics to the CFU-S found in normal and regenerating marrow and responds to CFU-S-specific regulators, to assay for the earliest appearance of 'stem cells' in differentiating EBs.

CFU-A (colony forming units type A) assay

The *in vitro* CFU-A assay was set up essentially as described previously (Pragnell *et al.*, 1994). Briefly:

1 A 1 ml feeder layer consisting of 6% Noble agar in αMEM, containing 10% donor horse serum and containing conditioned medium as a source of cytokines is established in a 3 cm diameter tissue culture Petri dish. The feeder layer contains 10% conditioned medium from the L-929 fibroblast cell line (a crude source of CSF-1 [ECACC85011425]) and 5% conditioned medium from the AF1-19T cell line (a crude source of GM-CSF. The AF1-19T cell line is an NRK rat fibroblast cell line transformed with a malignant histiocytosis sarcoma virus (Franz *et al.*, 1985)). These cell lines are grown in roller bottles with modified Eagles medium (MEM) and 10% foetal calf serum to half-confluence. Conditioned medium is removed, filtered and stored at −20 °C.
2 EBs either intact (50–150) or homogenised (see above – 10^5cells/plate) are added to 3% Noble agar in αMEM supplemented with 10% horse serum to form an upperlayer.
3 The dishes are incubated for 11 days in a humidified atmosphere of 10% CO_2 and 5%O_2.
4 In this assay CFU-A are scored as macroscopic colonies of >1 mm diameter (Fig. 1.3).
5 The self-renewal potential of these colonies may be checked by 'picking' them after 7 days of culture. After disaggregating by vigorous pipetting in 100 ml of αMEM, individual CFU-A cell suspensions are then re-plated in the CFU-A assay.

The CFU-A assay and others that can detect primitive multipotential haematopoietic progenitors are useful, quick and easy assays for the early detection of haematopoiesis *in vitro*. They are however unable to define the

(a)

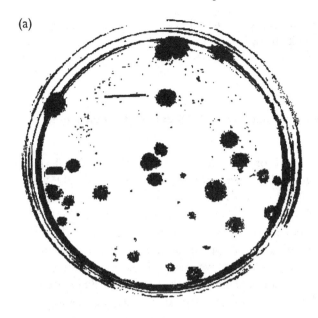

(b)

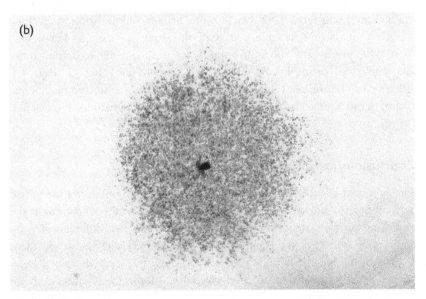

Fig. 1.3(a). Petri dish containing large CFU-A colonies clearly visible to the naked eye. (b). Close-up of a single CFU-A generated from an intact embryoid body. The remains of the three-dimensional structure of the body are at the centre of the colony.

true long-term repopulating HSCs. *In vitro* functional detection of lymphoid precursors is also difficult. In order to overcome these problems, adoptive transfer into irradiated or otherwise haematopoietically-compromised animals is necessary.

Reconstitution of recipient animals with ES-derived haematopoietic progenitors

Multilineage long-term reconstitution of recipients with ES-derived haematopoietic progenitors has been reported by Palacios *et al.* (1995) and ourselves (Hole *et al.*, 1996*a*). In the former paper, ES-cell to haemato-poietic progenitor cell differentiation was achieved in culture conditions using a stromal layer supplemented with an undefined cytokine cocktail. In our system, having identified the time-point of first appearance of haemat-opoietic cells using the CFU-A assay, we reasoned that repopulating cells may be present 24 hours earlier. Thus EBs after 4 days of differentiation (without added cytokines) were disaggregated by the Dispase method (see above) and injected (along with limiting doses of differentially marked carrier spleen cells) intravenously into mice irradiated with 10.5 Gy of γ-irradiation. Long-term (>2 years) multi-lineage repopulation with ES-derived progenitors was achieved with this strategy albeit at low levels (<10% of peripheral blood). Intriguingly we were also able to demonstrate that injection of embryoid body cell suspensions after 5 days of differentiation promoted long-term survival of the irradiated recipients without concomitant evidence of ES-cell derived repopulation (Hole *et al.*, 1996*a*).

Conclusions/prospectives

The potential of utilising totipotential stem cells as an unlimited source of haematopoietic stem cells (and possibly others) is only now becoming real-ised. The increasing definition of the programme of *in vitro* differentiation of embryonal stem cells and primordial germ cells will allow more detailed analysis of the haematopoietic process. Although the work described here refers only to embryonal stem cells derived from mouse, the isolation of these cells from other species, including man, would allow the development of new approaches to the detection and possible therapy of haematopoietic disease.

Acknowledgements

Work described in this chapter has been supported in part by grants from the Leukemia Research Fund, Wellcome Trust and Melville Trust for Cancer Research. The authors acknowledge the help and advice of their colleagues Dr Gerry Graham and Dr Austin Smith and the technical assistance of Janice Murray, Helen Taylor and Kay Samuel.

References

Chen, U., Kosco, M. & Staerz, U. (1992). Establishment and characterisation of lymphoid and myeloid mixed-cell populations from mouse late embryoid bodies, 'embryonic-stem-cell fetuses'. *Proc. Natl. Acad. Sci. USA*, **89**, 2541–5.

Doetschman, T., Eisetter, H., Katz, M., Schmidt, W. & Kemler, R. (1985). The in vitro development of blastocyst-derived embryonic stem cell lines: formation of visceral yolk sac, blood islands and myocardium. *J. Embryol. Exp. Morphol.*, **87**, 27–45.

Forrester, L. M., Bernstein, A., Rossant, J. & Nagy, A. (1991). Long-term reconstitution of the mouse hematopoietic system by embryonic stem cell-derived fetal liver. *Proc. Natl. Acad. Sci. (USA)*, **89**, 7514–17.

Franz. T., Lohler, J., Fusco, A., Pragnell, I., Nobis, P., Padua, R. & Ostertag, W. (1985). Transformation of mononuclear phagocytes in vivo and malignant histiocytosis caused by a novel murine spleen focusing virus. *Nature*, **315**, 149–51.

Hole, N. & Ansell, J. D. (1996). In vitro differentiation of embryonal stem cells as a tool for the study of haematopoiesis. In *Weir's Handbook of Experimental Immunology*, vol. IV, 5th edn, ed. L. A. Herzenberg, D. Weir, L. A. Herzenberg & C. C. Blackwell, pp. 147.1–147.5. Oxford: Blackwell Science.

Hole, N. & Smith, A. G. (1994). Embryonic stem cells and hematopoiesis. In *Culture of Hematopoietic Cells*, ed. R. I. Freshney, I. B. Pragnell & M. G. Freshney, pp. 235–49. New York: Wiley-Liss.

Hole, N., Graham. G. J., Menzel, U. & Ansell, J. D. (1996a). A limited temporal window for the derivation of multilineage repopulating progenitors during embryonal stem cell differentiation. *Blood*, **88**, 1266–76.

Hole, N., Leung, R., Doostdar, L., Menzel, U., Samuel, K., Murray, J., Taylor, H., Graham, G. & Ansell, J. D. (1996b). Haematopoietic differentiation of embryonal stem cells in vitro. In *Gene Technology*, ed. A. Zander, pp. 3–17. Springer Verlag.

Holyoake, T. L., Alcorn, M. J., Richmond, L., Farrell, E., Pearson, C., Green, R., Dunlop, D. J., Fitzsimons, E., Pragnell, I. B. & Franklin, I. M. (1997). CD34 + ve PBPC expanded ex vivo may not provide durable engraftment following myelo-ablative chemo-radiotherapy regimes. *Bone Marrow Transplant.*, **19**, 1095–101.

Hooper, M. L. (1994). *Embryonal Stem Cells*. Chur: Harwood Academic.

Keller, G., Kennedy M., Papayannopouloum, T. & Wiles, M. V. (1993). Hematopoietic commitment during embryonic stem cell differentiation in culture. *Mol. Cell. Biol.*, **13**, 473–86.

Nakano, T., Kodama, H. & Honjo, T. (1994). Generation of lymphohematopoietic cells from embryonic cells in culture. *Science*, **265**, 1098–101.

Nichols, J., Evans, E. P. Smith, A. G. (1990). Establishment of germ-line competent embryonic stem (ES) cells using differentiation inhibiting activity (DIA). *Development*, **110**, 1341–8.

Palacios, R., Golunski, E. & Samardis, J. (1995). In vitro generation of haematopoietic stem cells from an embryonic stem cell line. *Proc. Natl. Acad. Sci. USA*, **92**, 7530–4.

Pragnell, I. B., Freshney, M. G. & Wright, E. G. (1994). CFU-A assay for measurement of murine and human early progenitors. In *Culture of Haematopoietic Cells*, ed. R. I. Freshney, I. B. Pragnell & M. G. Freshney, pp. 67–79. New York: Wiley-Liss.

Schmidt, R. M., Bruyns, E. & Snodgrass, H. R. (1991). Hematopoietic development of embryonic stem cells in vitro: cytokine and receptor expression. *Genes Develop.*, **5**, 728–40.

Smith, A. G. (1991). Culture and differentiation of embryonic stem cells. *J. Tiss. Cult. Meth.*, **13**, 89–94.

Smith, A. G., Heath, J. K., Donaldson, D. D. *et al.* (1988). Inhibition of pluripotential embryonic stem cell differentiation by purified polypeptides. *Nature*, **336**, 688–90.

Wiles, M. V. (1993). Embryonic stem cell differentiation in vitro. *Meth. Enzymol.*, **225**, 900–18.

Wiles, M. V. & Keller, G. (1991). Multiple hematopoietic lineages develop from embryonic stem (ES) cells in culture. *Development* **111**, 259–67.

2

Dendritic cells

Jonathan M. Austyn, David Chao, Chen-Lung Lin, Justin A. Roake
and Rakesh Suri

Introduction

It is now well established that dendritic cells (DC) can function as specialised antigen-presenting cells for the initiation of immune responses against
foreign antigens. However, recent evidence indicates that there are developmentally and functionally distinct DC populations, and that one or more of
these subsets may play a critical role in the induction of thymic (central) and
perhaps extrathymic (peripheral) tolerance to self antigens. This chapter
focuses particularly on DC that are involved in the induction of immunity
to foreign antigens; increasing evidence suggests that these cells can be used
for immunotherapy of certain cancers and perhaps of microbial diseases. For
a review, containing sources of all uncited information in this chapter, see
Austyn (1998).

DC in non-lymphoid and lymphoid tissues

The ultimate fate of all DC is the lymphoid tissues. However, before DC can
induce immunity it is essential that they acquire foreign antigens that gain
access to peripheral tissues. Hence, after production in the bone marrow, DC
can enter non-lymphoid tissues, where they develop into cells with specialised capacities for antigen uptake and processing, and expression of foreign
peptide–MHC complexes at the cell surface. These cells can be termed
'immature' or 'processing' DC. DC are present in all epithelial sites, such as
skin epidermis, where they are referred to as Langerhans cells (LC), and the
mucosae of the respiratory, gastrointestinal, and urogenital tracts. They are
also present in interstitial spaces of vascularised organs such as heart and
kidney (Austyn, 1998).

Inflammatory cytokines and/or bacterial products such as lipopolysaccharide (LPS) induce dramatic phenotypic and functional changes ('maturation' or 'activation') of DC in non-lymphoid tissues and they migrate into secondary lymphoid tissues, via afferent lymph into lymph nodes and via blood to spleen. At the same time a wave of DC progenitors is recruited to the non-lymphoid site of inflammation or infection, presumably to repeat the cycle. DC can also be recruited to liver sinusoids before undergoing a blood–lymph translocation and migrating to the celiac lymph nodes. This route may be required for induction of immunity to blood-borne particulate antigens (Austyn, 1996).

An important characteristic of DC is their capacity to home to T cell areas of secondary lymphoid tissues, and to adhere in an antigen-independent manner to resting T cells presumably to facilitate subsequent antigen presentation. The fully 'mature' or 'costimulatory' DC can no longer internalise and process antigens, but they express high levels of foreign peptide–MHC complexes as well as costimulatory molecules such as CD40, CD80 and CD86, which are required for T cell activation during antigen presentation. Mature DC can also secrete IL-12, which promotes the development of activated T cells that secrete type 1 cytokine profiles, depending on the local environment.

Phenotypically distinct subsets of DC have been identified in secondary lymphoid tissues (see e.g. Grouard *et al.*, 1996; Kelsall & Strober, 1996; Steinman, Pack & Inaba, 1997; Vremec & Shortman, 1997). These include 'nests' of DC that interrupt the marginal zone of spleen (marginal zone DC; MZDC), and interdigitating cells (IDC) in periarteriolar sheaths, the T cell areas of central white pulp. These subsets probably correspond to DC subsets in lymph nodes that are detected close to the subcapsular space or deep within paracortical regions respectively, and to corresponding subsets in Peyer's patch. The former cells may represent DC that have recently been recruited from non-lymphoid tissues, prior to their movement into T cell areas and development into IDC. In addition, a subset of DC has been identified in germinal centres (GC) of human lymphoid tissues (GCDC), that is clearly distinct from follicular DC (FDC), which comprise a different, probably non-leukocytic lineage. It seems likely that GCDC interact with memory T cells and maintain the GC reaction, while FDC retain cell surface immune complexes and induce affinity maturation of B cells.

Development and function of DC subsets

Langerhans cells, myeloid DC and monocyte-derived DC

A major advance in the field has been the ability to grow DC in large numbers in culture (for a review, see Young & Steinman, 1996). For example, growth and differentiation of mouse DC can be induced by culture of bone marrow cells in granulocyte–macrophage colony stimulating factor (GM-CSF). These cells resemble immature, processing DC but they develop into costimulatory DC when they are cultured in tumour necrosis factor-alpha (TNFα) or lipopolysaccharide (LPS). It is also possible to grow mature DC from mouse bone marrow by culture in GM-CSF and interleukin (IL)-4, and yields can be increased by inclusion of Flt-3 ligand.

Studies in humans have demonstrated the existence of developmentally and functionally distinct DC subsets (for reviews, see Young & Steinman 1996; Austyn, 1998). CD34$^+$ DC progenitors are present in fetal liver, cord blood, bone marrow and adult peripheral blood. Growth and differentiation of DC can be induced by culture of these progenitors in GM-CSF and TNFα, and yields can be increased by inclusion of c-kit ligand (stem cell factor; SCF), which expands the progenitor pool. Distinct CD34$^+$ progenitors have been identified according to whether or not they express the cutaneous lymphocyte-associated antigen (CLA). CD34$^+$ CLA$^+$ progenitors give rise to CD1a$^+$ cells resembling Langerhans cells (LC) via a committed CD14$^-$ CD1a$^+$ precursor. In contrast CD34$^+$ CLA$^-$ progenitors give rise to CD1a$^+$ cells lacking the characteristic features of LC, via a bipotential CD14$^+$ CD1a$^-$ precursor that can alternatively differentiate into macrophages in the presence of M-CSF. The DC progeny of CD34$^+$ CLA$^-$ progenitors have therefore been termed 'myeloid DC'. Evidence for a distinct developmental pathway for LC has also come from studies of knockout mice: LC are absent from skin of mice lacking a functional TGF-beta gene but DC are present in their lymphoid tissues, and converse phenotypes are seen in mice lacking relB or Ikaros transcription factors (Austyn, 1998).

Culture of human peripheral blood monocytes in GM-CSF and IL-4 or IL-13 induces their development into DC in the absence of significant proliferation (Young & Steinman, 1996; Austyn, 1998). The cells resemble immature, processing DC but they are not terminally differentiated and can develop into macrophages when they are cultured in M-CSF. Further development of these cells into mature, costimulatory DC can be induced by culture in TNFα, LPS, or monocyte-conditioned medium (MCM; Bender et al., 1996; Romani et al., 1996a), which is produced by culture of monocytes on immobilised IgG. The active constituents of MCM are not yet well defined, but cytokines such as IL-6 and prostaglandin E2 seem to be involved in its function.

Lymphoid DC

In addition to LC, myeloid DC, and monocyte-derived DC (see above), a subset(s) of 'lymphoid DC' has also been identified in thymus and secondary lymphoid tissues of both mouse and human (for reviews see Shortman *et al.*, 1997; Austyn, 1998).

In human, a multipotent CD34$^+$ CD10$^+$ progenitor has been isolated from bone marrow and thymus that can generate DC and lymphoid cells (T, B and natural killer cells) but not myeloid cells. In addition, a subset of CD4$^+$ CD11c$^-$ cells has been identified in T cell areas of secondary lymphoid tissues (Grouard *et al.*, 1997). These cells were originally termed 'plasmacytoid T cells' but are now known to develop into DC after isolation and culture in IL-3 and CD40 ligand (in the absence of GM-CSF). Their localisation close to and within the high endothelial venules of tonsils suggests they may enter the tissue directly from the blood, in which a phenotypically similar subset of DC has been identified.

In mouse, an apparently multipotent CD4^{lo} progenitor has been isolated from thymus that can generate DC and lymphoid but not myeloid cells *in vivo*, and can produce DC after culture in the absence of GM-CSF (Shortman *et al.*, 1997). A downstream thymic progenitor resembling a pro-T cell has also been identified that can generate DC and T but not B cells. *In vivo*, the DC progeny of these progenitors all express CD8 as alpha–alpha homodimers. A substantial proportion of CD8α^+ DC can be isolated from mouse spleen and lymph nodes. These cells, like CD8$^-$ DC, have co-stimulatory activity for T cell responses *in vitro*. However, they appear to express a ligand for Fas and have negative regulatory properties. In particular, after T cell activation, they induce Fas-dependent apoptosis of CD4$^+$ T cells and limit IL-2 production by and proliferation of CD8$^+$ T cells in a Fas-independent manner. It has been suggested that the regulatory properties of these lymphoid DC subset could be involved in induction or maintenance of peripheral tolerance (Steinman *et al.*, 1997; Vremec & Shortman, 1997).

Characterisation of DC generated in culture

DC have been isolated from a wide variety of non-lymphoid and lymphoid tissues of human, mouse and other species (for details see Romani *et al.*, 1996*b*; Caux *et al.*, 1999, and references cited by Austyn, Liddington & MacPherson, 1997). For many purposes, a more convenient source is DC that can be grown in large numbers from progenitors in culture. For example, there is now considerable interest in generating these cells for DC-based immunotherapy (for review, see Austyn, 1998). However, before use in *in vitro* or *in vivo* studies, it is important to characterise the cells in some detail.

Phenotype

Cell phenotype can be established by conventional flow cytometric and immunocytochemical procedures. During culture from progenitors, commitment to the DC lineage is indicated by loss of expression of CD14 and c-fms in human and myeloid markers such as CD32 (Fc gamma RII) in human and mouse. Useful DC-restricted markers of the mature cells include expression of CD83, p55 (fascin) and S100 in human and CD11c and DEC-205 in mouse; these cells also express high levels of MHC class II molecules.

Function

DC at the immature, processing stage can internalise soluble tracers by pinocytosis and (actin-dependent) macropinocytosis, and phagocytose particulates such as latex microspheres and zymosan. These capacities are down-regulated during maturation. At the same time the cells acquire co-stimulatory activity and the ability to initiate T and T-dependent responses. The latter can be assessed in a variety of *in vitro* assays that are applicable to both human and mouse systems. These include oxidative mitogenesis and allogeneic mixed leukocyte responses. The former is a rapid, polyclonal proliferative response that occurs when DC are co-cultured with (syngeneic or allogeneic) T cells, after treatment of either population with sodium periodate; the magnitude of the response appears to correlate well with DC co-stimulatory activity. The latter is due to activation and proliferation of alloreactive ($CD4^+$ and $CD8^+$) T cells, typically purified from mouse lymph nodes or spleen, or from human peripheral blood (although human cord blood is preferable, if available, since it represents the best source of naive T cells).

Migration

In addition to their characteristic phenotypic and functional characteristics (see above), DC have specialised migratory properties (Austyn et al., 1996). In particular, mouse DC home to T cell areas of spleen and regional lymph nodes after administration via the blood or lymph respectively (the latter, for example, after footpad injection), and to the medulla of fetal thymus in organ culture. Frozen section assays have also been developed to demonstrate specific attachment of mouse DC to splenic marginal zone, presumably reflecting their point of entry to the tissue *in vivo*. Furthermore, human DC have been shown to undergo chemotaxis and transendothelial migration in response to defined chemokines. These properties can be assessed in trans-well systems *in vitro*.

Purpose of this chapter

The techniques decribed below have been selected for two reasons. First, they enable mouse and human DC to be generated in culture in relatively large numbers from perhaps the most readily available sources: mouse bone marrow cells and human peripheral blood monocytes. These approaches should permit immunobiological studies of DC to be performed more easily in the laboratory, and provide the basis for DC-based immunotherapy in the clinic. Second, these techniques permit the migratory properties of these cells to be assessed. These applications are discussed in the subsequent section: *in vivo* for mouse DC and *in vitro* for human DC. Investigation of the phenotype and *in vitro* functions of DC are relatively straightforward (see above), but studies of the equally important characteristic migratory proper-ties of DC are less commonly investigated. It is hoped that the procedures described below will provide a starting point for the novice and a conven-ient summary for the initiate.

Techniques

General requirements

Animals

Standard laboratory mice (>6 weeks old), e.g. C57BL/10 or BALB/c strains (e.g. Harlan-Olac, Bicester, Oxfordshire, UK).

Culture media and reagents

RPMI 1640, LPS-free (Gibco, Paisley, UK)
Fetal calf serum (FCS), low LPS (less than 0.2 ng/ml; Gibco); this should be
    heat inactivated by incubation at 56 °C for 30 min
Glutamine–penicillin–streptomycin × 100, LPS-free, Catalog No. 10378-
    016 (Gibco)
RPMI supplemented with 10% FCS plus glutamine and antibiotics (R10);
    complete culture medium
Phosphate buffered saline (PBS) (e.g. Oxoid, Unipath, Basingstoke, UK)
Tris buffered ammonium chloride (TBAC) for erythrocyte lysis
Trypan blue (e.g. Sigma, 0.6%) used at a 1 in 6 dilution to assess cell viabil-
    ity

Plastics

Standard sterile disposables, including pipettes (e.g. 1 ml, 5 ml, 10 ml, 25 ml), tubes (e.g. 1 ml, 5 ml, and 10 ml round-bottom tubes; 15 ml and 50 ml conical tubes), Petri dishes (e.g. 30 mm), syringes and needles.

Culture of mouse bone-marrow-derived DC

Immature, processing DC can be generated from mouse bone marrow cells cultured in GM-CSF provided that potential sources of LPS (endotoxin), including some laboratory water supplies, are excluded. Mature, costimulatory DC can be generated by culture in GM-CSF and subsequent exposure to TNFα or LPS, or by continuous culture in a combination of GM-CSF and IL-4. In our hands the latter procedure generates the most mature cells. However, it is most important to note that different results can be obtained in different laboratories, depending in part on variations between batches of fetal calf serum and, for example, loss of cytokine activity on repeated freezing and thawing of aliquots. The reader is referred to other publications for references to alternative or additional procedures (e.g. use of Flt-3 ligand) (Young & Steinman, 1996; Romani *et al.*, 1996*b*).

Specialised materials and equipment

Reagents

EDTA
Bovine serum albumin (BSA) (e.g. Sigma)
Recombinant cytokines. Lyophilised cytokines should be reconstituted in sterile, LPS-free PBS containing 0.1% bovine serum albumin and stored at 1 μg/ml at −80 °C until use: murine granulocyte–macrophage colony stimulating factor (GM-CSF; Peprotech, Rocky Hill, NJ); murine interleukin-4 (IL-4) (Peprotech, Rocky Hill, NJ); murine tumor necrosis factor-alpha (TNFα) (Peprotech, Rocky Hill, NJ).
Lipopolysaccharide (LPS) from *Salmonella typhimurium* (Difco Labs, Detroit, MI)

Plastics

50 ml conical tubes (polypropylene tubes can reduce cell losses, e.g. Falcon, #2063)
6-well tissue culture plates (Greiner, Germany)
10–30 ml syringes and sterile needles (19–26 gauge)
Nylon cell strainers (Falcon #2350)

Equipment

Scissors

Forceps

MiniMACS immunomagnetic bead depletion system (Miltenyi Biotec GmbH, Germany):

MiniMACS magnet and stand

Magnetic bead columns

Goat anti-rat Ig microbeads

Method

Obtaining bone marrow cells Animals are sacrificed by CO_2 asphyxiation and the lower limbs are cleansed with 70% ethanol solution. Femoral and tibial bones are dissected out in a sterile fashion and placed in sterile PBS for transfer to tissue culture hood. Bones are immersed in 70% ethanol for 3 min in a small Petri dish and then transferred to a dish containing RPMI.

In a sterile fashion, the epiphyses of bones are removed using scissors. A syringe with a fine (e.g. 21–26-gauge) needle containing RPMI is inserted into one end of each bone core and the marrow is expelled into a 50 ml conical tube. The needle is then withdrawn and inserted into the other end, and the bone is flushed again. Successful elution of the marrow core is evidenced by the loss of red hue to the bone shaft; usually only 1–2 ml RPMI need be used for each bone.

A single cell suspension is generated by resuspending the contents of the tube (e.g. by using a Pasteur pipette or by passing through a 19-gauge needle attached to a syringe). The suspension is filtered through a nylon cell strainer into another 50 ml conical tube to remove bone fragments. Cells are pelleted by centrifugation; all spins are at $250g$ for 7 min. Erythrocytes are lysed by resuspending in 5 ml TBAC for 5 min followed by quenching with at least 25 ml R10 before pelleting cells again.

Culture of (immature) bone marrow-derived dendritic cells Bone marrow cells are resuspended to a final concentration of 2×10^6 cells/ml in R10 medium supplemented with GM-CSF (10 ng/ml). Suspensions are plated at 4 ml/well in 6-well tissue culture plates and incubated for 2 days.

On day 2 post initiation of cultures, plates are swirled gently for 1 min and half of the existing medium is aspirated (together with non-adherent cells, predominantly granulocytes) and discarded. Fresh medium with the appropriate cytokine (10 ng/ml) is then added. We have operated upon the premise that cytokines are present in excess and have only replaced that removed in

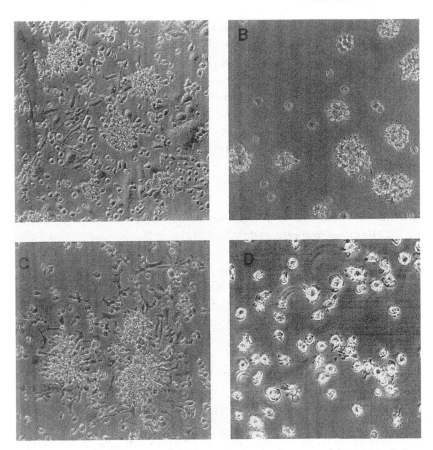

Fig. 2.1. Phase contrast micrographs illustrating the development of dendritic cells in mouse bone marrow cultures supplemented with GM-CSF. (A) Low power view of aggregates of developing dendritic cells attached to adherent cells at day 4. (B) Aggregates in suspension dislodged from marrow cultures at day 4. (C) After replating aggregates such as in (B), many cells reattach to tissue culture plastic by 3 h. (D) High power view of cells that release after further overnight incubation from cultures such as in (C). (Reprinted with permission from Inaba, K., Inaba, M., Romani, N., Aya, H., Degichi, M., Ikehara, S., Muramatsu, S. & Steinman, R. M. (1992). Generation of large numbers of dendritic cells from mouse bone marrow cultures supplemented with granulocyte/macrophage colony-stimulating factor. *J. Exp. Med.*, **176**, 1693–702.)

the medium renewal process. Most of the emergent granulocytes should be removed without dislodging the developing DC aggregates that are weakly tethered to a firmly attached layer of adherent cells until at least day 4 (e.g. Fig. 2.1A). Often, at this stage, non-adherent cells appear to have detached from the clusters and can be seen to have dendritic cell morphology.

At day 4 the contents of each culture well are split into two fresh wells to decrease cell density and each is refreshed with new medium and cytokine and cultured for a further 2 days. To do this, at day 4, the aggregates can be carefully dislodged from the underlying adherent cells, for example by Pasteur pipette (e.g. Fig. 2.1B). The cells tend to adhere within a few hours of plating into fresh wells (e.g. Fig. 2.1C) and to detach after overnight culture (e.g. Fig. 2.1D).

Harvesting and purification of dendritic cells By day 6, many non-adherent cells with pseudopodia-like processes should be present in the cultures (e.g. Fig. 2.1D). Plates are swirled gently to dislodge loosely adherent clusters of DC, and the non-adherent cells and medium are harvested. Typical bulk cell yield in our hands for two tibias and two femurs obtained from each C57BL/10 mouse is approximately $1.5–2 \times 10^7$ cells.

Bulk BMDC cultures are known to have contaminating fractions of T and B lymphocytes (as well as macrophages and granulocytes). The purity of DC obtained from these cultures can be increased by enriching for the fraction that is negative for these lineage markers using the MiniMACS system.

Technique

Note: all steps must be performed quickly at 4 °C.

1 Harvest cell suspensions, count in trypan blue, wash in cold PBS and spin at 250*g* for 7 min to pellet.
2 Resuspend pellet in rat anti-mouse primary monoclonal tissue culture supernatants: e.g. 500 µl of anti-B220 (anti-CD45; B cell-restricted isotype) and 500 µl of KT3 (anti-CD3).
3 Incubate on ice for 60 min.
4 Wash once in 5 ml R10, spin.
5 Wash twice in 5 ml ice cold PBS/FCS, spin.
6 Resuspend in 80 µl of PBS/FCS.
7 Add 20 µl goat anti-rat MiniMACS magnetic beads, incubate 15 min on ice.
8 Set up column in MiniMACS unit with a new 15 ml conical tube below to receive each fraction:
   attach flow resistor and remove plastic sheath
   pipette 500 µl buffer solution (PBS + 0.5% BSA + 5 mM EDTA) into column, allow to flow through, and discard
   pipette cell suspension into column, wash through with 500 µl buffer and retain this negatively selected fraction; purified DC

remove flow resistor, wash column with 500 μl buffer twice and retain as wash fraction (if required)

remove column from apparatus, place over a fresh tube, pipette 1 ml buffer into column, and flush out positive cells using plunger (if required)

9 Wash cells twice in R10 and count. The purified DC fraction can either be used directly or replated for maturation (see below). In our hands, this procedure typically results in the loss of 15–30% of starting cell numbers.

Maturation of dendritic cells Resultant DC can either be used immediately or replated for a further 3-day maturation period in TNFα or LPS. For the former, cells are cultured for a further 3 days at 2×10^6 cells/ml in R10 with the standard GM-CSF concentration (10 ng/ml) plus 10 ng/ml TNFα. For the latter, 1–50 ng/ml LPS is substituted for the TNFα.

Culture of murine bone marrow from day 0 with 10 ng/ml each of GM-CSF and IL-4, instead of GM-CSF alone (but otherwise using the protocol as described above), results in the generation of mature DC. These DC are generally larger, more granular and more 'dendritic' in morphology than those generated in GM-CSF alone; they exhibit higher levels of expression of costimulatory (e.g. CD40, CD80, CD86), MHC class II, and DC-specific (e.g. CD11c and DEC-205) molecules, and they have greated stimulatory activity in the oxidative mitogenesis assay than those cultured in GM-CSF alone. Additional culture of these cells with TNFα or LPS can result in a further increase in these parameters (R. Suri, in preparation).

The suggested concentrations of cytokines and LPS indicated above work well in our hands. It is however advantageous to perform checkerboard titrations of cytokines and LPS in order to determine optimal concentrations for maturation of DC, as assessed by phenotype and function for example, since these can vary widely between different laboratories.

Culture of human monocyte-derived DC

Immature, processing DC can be generated from human monocytes cultured in GM-CSF plus IL-4, which suppresses macrophage development. Further development and maturation of DC can be induced by subsequent exposure to TNFα, LPS, or monocyte-conditioned medium (for details of the latter see Bender *et al.*, 1996; Romani *et al.*, 1996*a*). The reader is referred to other publications for references to alternative or additional procedures (e.g. use of c-kit ligand, stem cell factor) (Grouard *et al.*, 1996; Romani *et al.*, 1996*b*; Caux *et al.*, 1999).

Specialised materials and equipment

Culture media and reagents

Heparin sodium (mucous) 5000 U/ml (CP Pharmaceuticals, Wrexham, UK)

Lymphoprep (Nycomed, Oslo, Norway)

Recombinant cytokines. Lyophilised cytokines should be reconstituted in sterile, LPS-free saline and stored at 10 µg/ml at $-80\,^{\circ}$C until use: human granulocyte–macrophage colony stimulating factor (GM-CSF), 4.44×10^6 IU/400 µg (Sandoz, Basel, Switzerland); human interleukin-4 (IL-4), typically 5×10^6 U/ml but check specific activity of each batch with the supplier (Peprotech, Rocky Hill, NJ); human tumour necrosis factor-alpha (R&D Systems, Abingdon, UK)

Lipopolysaccharide (LPS) from *Salmonella typhimurium* (Difco Labs, Detroit, MI)

Anti-human CD3 antibody as tissue culture supernatant

Anti-human CD19 antibody as tissue culture supernatant

Dynabeads M-450 coated with sheep anti-mouse antibody (Dynal, Oslo, Norway)

Commercial buffy coat (Blood Transfusion Service, Bristol, UK) or donor for fresh whole blood as the source of peripheral blood mononuclear cells (PBMC)

Virkon (Antec International, Sudbury, UK)

Plastics
All plastics need to be of tissue culture grade and sterile.
15 ml and 50 ml centrifuge tubes
Six-well plates
5 ml and 10 ml pipettes
Syringes and sterile needles

Equipment
Dynal MPC-1 magnet (Dynal, Oslo, Norway)
Rotator in cold room

Method

Obtaining peripheral blood mononuclear cells (PBMC) There are two sources of blood as the starting material. The maximum yield of PBMC will come from leukaphoresis and is available commercially as buffy coat preparations from blood transfusion services. These should be no older than 48 h as cell yields fall rapidly after that. Each buffy coat preparation will typically yield $2–3 \times 10^9$ cells. Alternatively donors may be bled but the amounts of whole blood that can be taken are generally limited to 100–200 ml at a time. Each 50 ml of whole blood typically yields between $1–2 \times 10^7$ cells.

For commercial buffy coats, the blood is removed from the bag with a needle and syringe under sterile conditions. The volume of blood is noted for future reference. Dilute the blood 1:2 with RPMI and add 50 U heparin for each 100 ml of mixture. For fresh whole blood, add 100 U heparin per 50 ml syringe before taking the blood; the blood is diluted 1:2 as above but no further heparin is added.

Dispense 15 ml lymphoprep per 50 ml tube and gently layer on 30 ml of the mixture per tube taking care to preserve the interface. Centrifuge the tubes at $800g$ for 20 min at room temperature for fresh blood, or for 30 min for commercial buffy coats. After centrifugation note the volume of the red cell pellet for future reference.

Carefully remove the interface using a pipette and dilute the PBMC at least 1:5 with cold RPMI to reduce the density of the solution. Pellet cells by spinning at $500g$ for 7 min at $4\,°C$. Resuspend the cells with cold RPMI and spin again at $250g$. Repeat once more, spinning at $200g$. These lower speed spins reduce the platelet contamination. Do a viability cell count with trypan blue just before the last spin.

All blood should be treated as infectious material. Gloves should be used at all times. Waste blood products should be decontaminated in 1% Virkon overnight and soiled plastics should be autoclaved.

Culturing the monocyte-derived dendritic cells Our practice is to culture dendritic cells in either RPMI + 10% FCS or 1% autologous plasma, supplemented with glutamine and antibiotics. To use 1% plasma, remove the medium above the interface. The concentration of the plasma in this medium can be calculated from (total blood volume − red cell pellet volume)/(total blood volume − red cell pellet volume + volume of added RPMI). The plasma medium is filtered to remove platelets and aliquots stored at $-20\,°C$.

Resuspend the PBMC after the final wash in the chosen culture medium to a final cell concentration of 1×10^7 cells/ml and pipette 2 ml per well into six-well plates. Place in incubator for 2 h. To remove the non-adherent cells, swirl the plates vigorously and pipette the medium from each well. Gently run 2 ml of medium down the sides of the well, swirl and remove. Replace with 2 ml of medium. Add GM-CSF and IL-4 to each well to give 50 µg/ml final

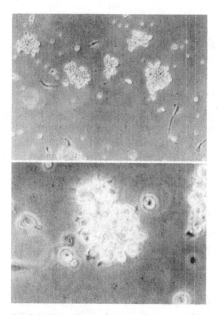

Fig. 2.2. Phase contrast micrographs of aggregates of developing dendritic cells in day 6 human monocyte cultures supplemented with GM-CSF and IL-4 at low (top) and higher (bottom) power.

concentration. Return to incubator and feed every other day by gently replacing 1 ml medium and adding a further 100 μl of GM-CSF and IL-4 per well.

In terms of specific cytokine activity our protocol equates to GM-CSF 500 U/ml and IL-4 250 U/ml, which is less than some protocols. We have found no benefit in increasing the cytokine concentrations and we suspect that our re-feeding schedule, which assumes complete exhaustion of all the cytokines, results in concentrations well above these figures. It is however advantageous to perform checkerboard titrations of cytokines (and LPS; see Maturation of dendritic cells, below) in order to determine optimal concentrations for maturation of DC, since these can vary between laboratories.

Harvesting and purification of dendritic cells Day 6 is the earliest that we use dendritic cells. Check under the inverted phase microscope that there are clusters of cells (e.g. Fig. 2.2). To harvest the cells wash vigorously over the surface of the well and keep the pooled medium on ice. Do a differential cell count by counting large dendritic cells and smaller lymphocytes separately. Typically a plate yields $2–3 \times 10^6$ dendritic cells with 20–50% contamination with lymphocytes. If a purer population of dendritic cells is required then proceed to lymphocyte depletion by magnetic beads. Note that losses of dendritic cells can be as high as 50% although more typically are around 10–20%. The most important factor is keeping everything at 4°C to prevent the dendritic cells from phagocytosing the Dynabeads. Note that plasma-cultured dendritic cells are highly adherent and should be kept on ice to prevent non-specific losses.

Pellet the cells at 250g for 7 min at 4°C. Resuspend the cells in sterile anti-CD3 (for T cells) and anti-CD19 (for B cells) tissue culture supernatant using 50 µl of each per 10^6 lymphocytes as ascertained by haemocytometer and incubate on ice for 30 min. Prepare Dynabeads following manufacturer's instructions. Briefly, a bead:lymphocyte ratio of at least 4:1 is needed. Wash the required number of beads in medium, resuspend in enough medium to give a concentration of at least 10^7 beads/ml and keep ice-cold. Wash the cells twice with ice-cold medium and resuspend in the bead solution. Transfer on ice to a cold room and leave on rotator for 20 min. Place a 15 ml tube in the MPC-1 magnet and add 10 ml of ice-cold medium. Add the bead and cell mixture and leave for 2 min. Pour off the non-magnetic dendritic cell fraction into a clean tube and do a differential cell count to check on the efficiency of the depletion. Typically lymphocyte contamination should fall to less than 5%. Wash the dendritic cells and resuspend in desired volume for use.

Maturation of dendritic cells Dendritic cells are resuspended to 10^6 cells/ml in medium containing GM-CSF and IL-4 and 2 ml aliquoted into each well of a six-well plate. To mature these cells further we add either 50 ng/ml LPS or 20 ng/ml TNFα. The cells are routinely harvested after 3 days for use although many maturation characteristics appear earlier. Monocyte-conditioned medium (MCM; Bender *et al.*, 1996; Romani *et al.*, 1996a) has also been shown to mature dendritic cells.

Applications

Migration of Tc-99m labelled mouse DC *in vivo*

A comprehensive overview of techniques to trace migration of DC in mice, including cell labelling with indium-111 or fluorochromes (H33342, DiI, PKH26, carboxyfluorescein), together with immunocytochemical techniques for colocalising the cells within recipient lymphoid tissues, is in the reference Austyn *et al.* (1997). Techniques are described below for tracing *in vivo* migration of DC labelled with Tc-99m (this section) and the fluorochrome CFDASE (below). In our hands, Tc-99m is of similar value to indium-111 but it may be more readily available in certain clinical centres. CFDASE has certain advantages over other fluorochromes, for example, it is relatively resistant to acetone fixation. Other fluorochromes can be used, but recent evidence suggests that labelling with carboxyfluorescein can damage DC or otherwise alter their migration from the blood (R. Suri *et al.*, in preparation).

Specialised materials and equipment

Radioisotope and carrier

99m-Technetium (Tc-99m, Amersham Healthcare, UK). Tc-99m disintegrates with the emission of gamma irradiation and has a half-life of 6.02 h

Ceretec kit (Amersham radiochemicals) including carrier Exametrazine (HM-PAO)

Plastics

50 ml conical tubes ('opaque' polypropylene tubes, which can reduce cell losses, such as Falcon, #2063)

Sterile 1 ml syringes and needles (30.5 gauge)

Equipment

A microbalance is required to weigh tissues

A gamma-counter to measure radioactivity

Methods

Technetium labelling of DC Due to the short half-life of Tc-99m, the isotope must be eluted from the source generator immediately before use. To label leukocytes it is coupled to the carrier agent Exametrazine (HM-PAO). The resultant Tc-HM-PAO complex must be used within 15–30 min of coupling. We obtain a sample of this complex from our Radiology Department (John Radcliffe Hospital) immediately after formulation for use in patient procedures; simple rinsing of the tube to be discarded provides more than sufficient activity for experiments.

To each vial containing 4.5 mg Ceretec (HM-PAO), add 5 ml of sterile eluent from a Tc-99M generator and mix thoroughly. Within 30 min of reconstitution, take an aliquot and dilute to 150 MBq/ml. Add 0.1 ml of this solution (15 MBq) to a cell pellet (2×10^6 DC) and resuspend the cells. Incubate the suspension for 15 min at 20 °C, wash twice in R10, and resuspend in the same for injection.

The activity of small aliquots of the wash supernatants and the final cell suspension should be measured in a gamma counter. The labelling efficiency can then be calculated as (total activity in cells) / (total activity in cells + total activity in supernatants). Labelling efficiency in our hands is routinely 60–80% and cell viability is unaltered by the procedure (as assessed by trypan blue exclusion). It is also useful to calculate the activity incorporated per cell, since it is then possible to extrapolate the approximate number of cells detected in tissues after migration (although free label can also accumulate).

Administration of Tc-99m-labelled dendritic cells Inject Tc-99m-labelled cells in R10 into anaesthetised recipients, for example intravenously (up to 0.5 ml via the penile or jugular vein) or subcutaneously (25 μl into one footpad). In our hands 1×10^5 cells labelled in the aforementioned manner is sufficient to quantify tissue radioactivity 24 hours later in a standard 25 g mouse. Retain all plastics and used syringes for disposal in appropriate radioactive waste facilities.

Measurement of radioactivity by tissue gamma-counting At various time intervals (such as 3 h, 24 h) after cell transfer, the animal is sacrificed. Tissues are removed and transferred to capped tubes that are suitable for use in a gamma counter. Weigh the tubes before and after addition of tissues, and calculate the weight of each tissue. Measure the activity of Tc-99m in each tissue. Weigh and count the eviscerated carcass to assist in calculation of total recovered label. Organ activity can then be expressed in two ways: (a) as a percentage of the total radioactivity injected (percent recovered); and (b) as a percentage of the total radioactivity recovered per gram tissue sample (specific recovered activity – SRA). SRA is a useful figure in that it allows for the difference in size of the tissues and the actual amount of label that is accessible. Theoretically, it should be possible to compare SRA values from various animals, as the results are controlled for injection efficiency. For these calculations it is advantageous to use a computer spreadsheet (e.g. Microsoft Excel).

Control preparations To establish that label has accumulated in tissues as a result of cell migration, it is important to carry out a number of control experiments. While none is ideal, they include injection of labelled cells that have been formaldehyde fixed, injection of sonicates of a similar number of labelled cells, and injection of free isotope.

Measurement of radioactivity by gamma camera It is possible to follow the course of DC migration over time by scanning the body of an anaesthetised animal that has received Tc-labelled DC. Resolution and sensitivity depend on the actual camera used. In our hands, resolution up to 2 mm is obtainable up to 24 h after injection of 1×10^7 Tc-labelled DC. Accumulation is highest in spleen and liver, which concurs with results from tissue counting.

Migration of fluorochrome-labelled mouse DC *in vivo*

Specialised materials and equipment

Culture media and reagents

Fluorochrome CFDASE: 5- and 6-carboxyfluorescein diacetate succinimidyl ester (C-1157, Molecular Probes, Eugene, Oregon, USA)
Hanks balanced salt solution (HBSS)
Dimethylsulphoxide (DMSO)
TissueTek (Miles Diagnostics, IL, USA) embedding compound
Acetone or 1% formaldehyde for fixation of tissues
Washing solution (WS: PBS+5% FCS+0.1% sodium azide for cell labelling for histochemistry or FACS)
Monoclonal antibodies (anti-mouse antigens) and relevant anti-Ig reagents
Substrates for subsequent cell localisation
Vectashield (Vector Labs) for mounting coverslips

Equipment

Microscopes. An ultraviolet microscope with filters suitable for detecting CFDASE (green) and contrasting (e.g. red) fluorescence is essential. An incident light microscope is important for immunocytochemical procedures
Cryostat. Required for preparation of 5–10 μm frozen sections on microscope slides
Four-spot or 12-spot coated slides (such as C. A. Hendley Ltd., Essex, UK) are useful for subsequent labelling of sections
Liquid N_2 and insulated Thermos for freezing of tissues
Storage freezers: $-80\,^\circ$C for tissue specimens and $-20\,^\circ$C for cut sections (wrapped in aluminium foil)

Methods

Fluorochrome labelling of dendritic cells For CFDASE labelling, prepare stock solution at 5 mM in DMSO and store at $-80\,^\circ$C in the dark. Prepare cells for labelling by washing once in warm RPMI without FCS. Dilute the stock to a 5 μM solution in warm RPMI, and add an appropriate volume to the cell pellet in order to resuspend the cells to $2\text{--}10\times10^6$/ml. Incubate in a water bath at 37 °C, shaking the tube every five minutes. Wash the cells in ice-cold R10 twice to stop the labelling.

Administration of fluorochrome-labelled cells Administer the cells as for Tc-labelled DC. At required times after cell administration (such as 3 h, 24 h) remove tissues (such as spleen, popliteal lymph nodes) from recipient mice, embed in TissueTek, and snap freeze in liquid N_2. Prepare frozen sections (e.g. 5–10 μm) immediately, or after storage of tissues at -80 °C. View under a UV microscope with appropriate filters.

Co-localisation of fluorochrome-labelled cells by UV microscopy It is possible to localise the fluorochrome-labelled DC in relation to distinct microenvironments within the spleen and nodes using leukocyte lineage markers conjugated to different color fluorochromes (e.g. Austyn *et al.*, 1977). Many different protocols are possible with the noted constraint that most fluorochromes which have been used to label DC are fixation sensitive (e.g. PKH-26 and CF). CFDASE is one of the few exceptions we have found to be relatively resistant to acetone fixation. A fixation-resistant dye should be used to label DC when further histological co-localisation is required. A general protocol for this is as follows.

1 Fix sections in acetone for 10 min. For 12-spot slides, a volume of 50 μl/spot is adequate here and below.
2 Rehydrate sections in WS + 20% mouse serum for 10 min.
3 Incubate with saturating concentrations of rat anti-mouse mAbs for 30 to 45 min at room temperature.
4 Wash five times with WS.
5 Incubate with saturating concentrations of e.g. red fluorescing TRITC-mouse anti-rat Ig (or comparable item) for 30 to 45 min at room temperature.
6 Wash five times with WS.
7 Mount in Vectashield with coverslip.
8 The sections can then be viewed under a UV microscope, and photographed if required.

Notes and recommendations In our hands, it is necessary to inject at least $2\text{--}4\times10^6$ labelled BMDC to detect sufficient cells in the spleen for co-localisation. The mountant is important to prevent fluorochrome leeching and bleaching.

To establish that label has accumulated in tissues as a result of DC migration, it is important to carry out a number of control experiments, as detailed above for radiolabel experiments.

Detection of fluorochrome-labelled cells by FACS analysis of spleen In order to address whether fluorochrome-labelled cells detected by UV microscopy are likely to be the same DC as were injected, it is useful to perform recovery experiments. Briefly, the spleen, for example, is removed from an animal that has been injected intravenously (i.v.) with at least 1×10^7 CFDASE-labelled DC and is processed in the following manner alongside a control spleen from a naive untreated mouse.

1 Remove spleen and inject 1 ml of a warm RPMI-based 4 mg/ml colla- genase-IV solution into the spleen parenchyma along the long axis of the organ causing 'ballooning'.
2 Incubate for 15 min at 37°C agitating every 5 min.
3 Tease apart spleen using forceps in a nylon mesh tea strainer which is resting in a Petri dish with 2 ml warm RPMI covering the bottom.
4 Use a syringe plunger to force all teased splenic tissue through the mesh.
5 Transfer cell suspension to a 15 ml conical tube. Discard any large undis- solved fragments.
6 Pellet cells and lyse erythrocytes as above.
7 Resuspend and wash twice in WS.
8 Aliquot 10^5 cells into FACS tubes and incubate for 30 min with anti- mouse monoclonal antibodies conjugated to a color other than green (e.g. conjugated to phycoerythrin or CY5), then wash twice with WS.
9 Analyse the cells by flow cytometry, gating on CFDASE (green / FL1) positive cells, and examine their co-expression of other markers.

The investigators have found CFDASE positive cells recovered from spleens 24 h after i.v. administration to be negative for B220 (B-cell), KT3 (T-cell), and F4/80 (macrophage) but approximately 50–60% positive for the DC restricted marker NLDC-145 (R. Suri, in preparation). This is taken to be provisional evidence that the CFDASE dye has remained with DC and decreases the likelihood that the label has leached out onto surrounding lym- phocytes or that the DC have been engulfed by splenic macrophages. It does not by itself, however, exclude uptake by other DC of similar phenotype.

Chemotaxis of human monocyte-derived DC in culture

Chemotaxis of DC generated from human monocytes by culture in GM- CSF and IL-4 or IL-13 has been studied in transwell systems, and the response to chemokines has received particular attention (for reviews on chemokines see Schall & Bacon, 1994 and Nelson & Krensky, 1998). DC have been shown to undergo chemotaxis in response to classical chemo-attractants such as formyl peptides and C5a, and a characteristic set of CC chemokines including MIP-1α, MIP-1β, and RANTES (which bind CCR5), MCP-3, and MDC (Austyn, 1998). Maturation of the cells in LPS, TNFα or IL-1β results in abolition of responses to the CC chemokines whereas responses to the CXC chemokine SDF-1 (which binds CXCR4) are maintained or dramatically enhanced (Lin et al., 1999). DC expression of CCR5 and CXCR4 is of particular interest since these molecules are co-receptors for HIV-1 and are required for entry of the virus to the cells (Granelli-Piperno et al., 1996; Moore, Trkola & Dragic, 1997). DC exhibit exquisite sensitivity to LPS so it is essential that contamination by endotoxin is excluded from all cultures and materials used for techniques described below.

Specialised materials and equipment

Culture media and reagents

Recombinant human chemokines (CXC, CC, and C chemokines) (e.g. from R&D Systems Europe), as required

Plastics

Transwell apparatus (see Fig. 2.3). For chemotaxis assays of DC we use 6.5 mm transwell (TW) inserts (Costar, Cambridge, MA) in standard 24-well plate wells (Greiner, UK). The transwell has a polycarbonate multi-porous membrane (10 µm thick with 5 µm pores) separating the transwell system into an upper and a lower chamber. DC are placed in the upper chamber and a source of chemoattractants or chemokines is placed in the lower chamber. Since the average diameter of a human monocyte-derived DC may be about 10 µm, high background due to non-specific falling through the 5 µm pores can be prevented.

Method

DC migration in the transwell system

1 Harvest the GM-CSF and IL-4 cultured human monocyte-derived dendritic cells from the six-well plates on day 5 to day 8. Wash with R10 twice to remove the cytokines or growth factors in the culture, and resuspend in the same to 5×10^5/ml.

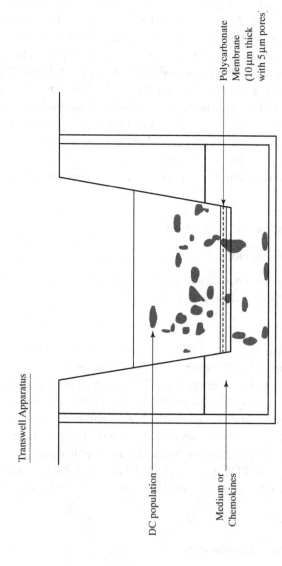

Transwell Apparatus

Polycarbonate
Membrane
(10 µm thick
with 5 µm pores)

DC population

Medium or
Chemokines

Fig. 2.3. Schematic illustration of a transwell apparatus.

2 Put chemoattractants or chemokines in the standard 24-well plate (i.e. the lower chamber of the TW system), e.g. at 50 ng/ml in 600 µl ordinary medium (RPMI or DMEM).

3 Add 100 µl cell suspension (containing 5×10^4 DC) into each TW insert, and place the TW inserts into the 24-well plate, which has been filled with 600 µl medium or chemokines, to create a two-chambered chemotactic assay model. Be sure that there is no air bubble below the polycarbonate membrane before culture. Place the TW apparatus in a 37 °C humidified CO_2 incubator for 2 h.

4 After 2 h incubation, lift the transwell inserts, and wash the bases of the inserts with 200 µl medium twice to dislodge the cells that migrated through the multiporous polycarbonate membrane but did not drop into the lower chamber.

5 Collect the cells in the lower chamber and transfer them into 5 ml U-bottom clear tubes.

6 Fix the cells with 5% formalin.

Quantification of the numbers of migrated DC Flow cytometry can be used to quantify the numbers of DC that migrate through the multi-porous poly-carbonate membrane in response to chemoattractants such as chemokines. Fluorescent beads (CC3F/PE/PE-CY5 beads, 7.6 µm in diameter, Flow Cytometry Standards Corp., San Juan, PR) are added into the tubes as an internal control (typically 5×10^4 beads per tube). In two-colour analysis, these beads are evident as a separate population from the DC. The acquisition of the cells is stopped after a certain number of the beads (1000 beads, for example) are acquired. The relative cell number of DC in each tube can be obtained by gating the large, granular cell population in the forward and side-scatter (FSC-SSC) dot plot (these parameters are related to cell size and granularity).

Transendothelial migration of human DC *in vitro*

A system to study transendothelial migration (TEM) of DC in transwell systems has been developed (Rahdon *et al.*, 1997) since these assays may be more relevant to *in vivo* migration than is simple chemotaxis in 'bare' transwell systems as described above. Using techniques described below, we have demonstrated that DC generated from monocytes cultured in GM-CSF plus IL-4 undergo TEM in response to the CC chemokines MIP-1α, MIP-1β, RANTES and MCP-3; maturation of the cells in LPS, TNFα or IL-1β

abolishes these responses but enhances the TEM response to SDF-1 (Lin *et al.*, 1999).

Specialised materials and equipment

Cells, culture media and reagents

Human microvascular endothelial cell line-1 (HMEC-1). An immortalised human endothelial cell line, isolated from neonatal foreskin, transfected with SV-40, cultured and passaged in MCDB-131 medium (below). For our experiments, the HMEC-1 line was kindly provided by Center of Disease Control, Atlanta, Georgia, USA

Culture medium. MCDB-131 (Sigma, St Louis, MO) containing 20% FCS, 10 ng/ml epidermal growth factor (EGF; Sigma), 1 µg/ml hydrocortisone (Sigma), antibiotics and L-glutamine (Gibco, Paisley, UK)

Fibronectin, isolated from human plasma (Cell Biology Boehringer Mannheim, Germany)

^{14}C-mannitol (Amersham, UK)

Scintillant fluid (Optiphase Hisafe, Fisons, UK)

Plastics

Transwell apparatus.

Equipment

Beta-radioactivity counter (e.g. 1219 Rackbeta, LKB, Wallac)

Method

Coating TW polycarbonate membranes with fibronectin The upper surface of the membrane is incubated with human fibronectin at a concentration of 5 µg/cm^2 at 37 °C in 5% CO_2 for at least 1 h; it is not necessary to place any medium in the lower chamber since fluid does not leak from the upper chamber. At the end of the incubation, the fibronectin is gently aspirated without touching the membrane. It is not necessary to wash the membrane with PBS or other medium.

Growth of HMEC-1 on the fibronectin-coated membrane The HMEC-1 cells are detached from a confluent T75 flask after treatment with 1 mM EDTA/PBS. The cells are resuspended in MCDB-131 to 5×10^5/ml. 100 µl of the cell suspension is placed in the upper chamber of the transwell apparatus, and 600 μl of MCDB-131 is placed in the lower chamber. Again, be sure there is no air bubble below the polycarbonate membrane after the transwell system is set up. Culture the cells in the transwell apparatus at 37 °C in 5% CO_2 for 3 days.

Integrity test After endothelial cells (EC) have been cultured on transwells for 3 days, 4.4×10^4 d.p.m. (disintegrations per min; 1 μ Ci is 2.22×10^6 d.p.m.; 1 Bq is 60 d.p.m) of ^{14}C-mannitol in 10 μl PBS is added to 100 μl of MCDB-131 in each TW insert and incubated for 30 min at 37 °C in 5% CO_2. At the end of the incubation period 50% of the medium in both the upper $((100 + 10)/2 = 55$ μl) and lower chamber $(600/2 = 300$ μl) is removed and placed in opaque scintillant tubes with 5 ml of scintillant. Activity is measured using a beta-counter. The results are expressed as a concentration ratio (CR) of the ^{14}C-mannitol in the upper chamber to that in the lower chamber. A CR of >20 is considered indicative of a satisfactory endothelial monolayer.

Transendothelial migration assay of DC DC are cultured and prepared as described above and aliquoted at 2×10^5 in 100 μl RPMI 1640 containing 10% FCS, placed into the upper compartment, and allowed to migrate through the confluent HMEC-1 monolayer to the bottom well, which contains media with various concentration of chemokines (e.g. 50 ng/ml) or control media. The transwells are incubated at 37 °C, 5% CO_2 for 4 h, then each insert is removed from its well and the base is rinsed twice with 200 μl medium to dislodge the cells that have migrated through but not sedimented to the bottom well. The cells are harvested from the bottom wells, transferred to 5 ml U-bottomed clear tubes, and fixed with 5% formalin.

Quantification of the numbers of transmigrated DC See above.

References

Austyn, J. M. (1996). New insights into the mobilization and phagocytic activity of dendritic cells [comment]. *J. Exp. Med.*, 1996, **183**, 1287–92.

Austyn, J. M. (1998). Dendritic cells. *Curr. Opin. Hematol.*, **9**, 1–15.

Austyn, J. M., Liddington, M. A. & MacPherson, G.G. (1996). Dendritic cells: migration in vivo. In *Weir's Handbook of Experimental Immunology*, 5th edn, ed. L. A. Herzenberg, D. M. Weir, L. A. Herzenberg & C. Blackwell, Oxford: Blackwell Science.

Bender, A., Sapp, M., Schuler, G., Steinman, R. M. & Bhardwaj, N. (1996). Improved methods for the generation of dendritic cells from nonproliferating progenitors in human blood. *J. Immunol. Meth.*, **196**, 121–35.

Caux, C., Dezutter-Dambuyant, C., Liu, Y.-J. & Banchereau, J. (1999). Isolation and propagation of human dendritic cells. In *Methods in Microbiology: Immunological Methods*, ed. D. Kabelitz & K. Ziegler. Academic Press.

Granelli-Piperno, A., Moser, B., Pope, M., Chen, D., Wei, Y., Isdell, F., O'Doherty, U., Paxton, W., Koup, R., Mojsov, S., Bhardwaj, N., Clark, L. I., Baggiolini, M. & Steinman, R. M. (1996). Efficient interaction of HIV-1 with purified dendritic cells via multiple chemokine coreceptors. *J. Exp. Med.*, **184**, 2433–8.

Grouard, G., Durand, I., Filgueira, L., Banchereau, J. & Liu, Y. J. (1996). Dendritic cells capable of stimulating T cells in germinal centres. *Nature*, **384**, 364–7.

Grouard, G., Rissoan, M. C., Filgueira, L., Durand, I., Banchereau, J. & Liu, Y. J. (1997). The enigmatic plasmacytoid T cells develop into dendritic cells with interleukin (IL)-3 and CD40-ligand. *J. Exp. Med.*, **185**, 1101–11.

Kelsall, B. L. & Strober, W. (1996). Distinct populations of dendritic cells are present in the subepithelial dome and T cell regions of the murine Peyer's patch. *J. Exp. Med.*, **183**, 237–47.

Lin, C.-L., Suri, R. M., Rahdon, R. A., Austyn, J. M. & Roake, J. A. (1999). Dendritic cell chemotaxis and transendothelial migration are induced by distinct chemokines and is regulated on maturation. *Eur. J. Immunol.*, **28**, 4114–22.

Moore, J. P., Trkola, A. & Dragic, T. (1997). Co-receptors for HIV-1 entry. *Curr. Opin. Immunol.*, **9**, 551–62.

Nelson, P. J. & Krensky, A. M. (1998). Chemokines, lymphocytes and viruses: what goes around, comes around. *Curr. Opin. Hematol.*, **10**, 265–270.

Rahdon, R. A., Lin, C. L., Suri, R. M., Morris, P. J., Austyn, J. M. & Roake, J. A. (1997). An endothelial cell-derived chemotactic factor promotes transendothelial migration of human dendritic cells. *Transplant Proc.*, **2**, 1121–2.

Romani, N., Reider, D., Heuer, M., Ebner, S., Kampgen, E., Eibl, B., Niederwieser, D. & Schuler, G. (1996a). Generation of mature dendritic cells from human blood. An improved method with special regard to clinical applicability. *J. Immunol. Meth.*, **196**, 137–51.

Romani, N., Bhardwaj, N., Pope, M., Koch, F., Swiggard, W. J., O'Doherty, U., Witmer-Pack, M. D., Hoffman, L., Schuler, G., Inaba, K. & Steinman, R. M. (1996b). Dendritic cells. In *Weir's Handbook of Experimental Immunology*, 5th edn, ed. L. A. Herzenberg, D. M. Weir, L. A. Herzenberg & C. Blackwell. Oxford: Blackwell Science.

Schall, T. J. & Bacon, K. B. (1994). Chemokines, leukocyte trafficking, and inflammation. *Curr. Opin. Immunol.*, **6**, 865–73.

Shortman, K., Wu, L., Suss, G., Kronin, V., Winkel, K., Saunders, D. & Vremec, D. (1997). Dendritic cells and T lymphocytes: developmental and functional interactions. *Ciba Found. Symp.*, **204**, 130–8.

Steinman, R. M., Pack, M. & Inaba, K. (1997). Dendritic cells in the T-cell areas of lymphoid organs. *Immunol. Rev.*, **156**, 25–37.

Vremec, D. & Shortman, K. (1997). Dendritic cell subtypes in mouse lymphoid organs. Cross-correlation of surface markers, changes with incubation, and differences among thymus, spleen, and lymph nodes. *J. Immunol.*, **159**, 565–73.

Young, J. W. & Steinman, R. M. (1996). The hematopoietic development of dendritic cells: a distinct pathway for myeloid differentiation. *Stem Cells*, **14**, 376–87.

3

Murine thymic explant cultures

Kim. L. Anderson and John. J. T. Owen

Cell culture systems allow the investigation of particular stages of T cell development under defined *in vitro* conditions. Of the systems currently in use for studying the very early stages of thymocyte development, thymic explant cultures provide the only system capable of supporting a complete programme of thymocyte development *in vitro*. Whole lobe cultures provide the opportunity to follow the progress of T cell development *in vitro* in the absence of further colonisation and have been widely used to examine the effects of a number of treatments on T cell development (see Jenkinson & Anderson, 1994 for review). Fetal thymus organ cultures have also been used to study the role of peptides in positive selection (Ashton-Rickardt et al., 1993, Hogquist, Gavin & Bevan, 1993), the mechanism of the $CD4^-8^-$ to $CD4^+8^+$ transition (Levelt, Ehrfeld & Eichmann, 1993a; Levelt et al., 1993b) and more recently for retroviral transfection studies on the signalling pathways involved in the double negative to double positive transition (Crompton, Gilmour & Owen, 1996). However, although un-manipulated thymus lobes are useful for studying some aspects of T cell development and the signalling pathways involved in apoptosis, the intrinsic heterogeneity of these lobes proves limiting for studies on the interactions between thymocytes and individual stromal cell types. Techniques have therefore been developed that allow the association of defined stromal and lymphoid populations in 'reaggregate' thymus organ cultures where optimal three-dimensional conditions are maintained (Jenkinson, Anderson & Owen, 1992). This approach has been used to establish chimaeric thymus lobes *in vitro* (Merkenschlager & Fischer, 1994) and to study the role of individual stromal cell types in the selection of the T cell repertoire (Anderson et al., 1993).

Introduction

The majority of T lymphocytes that bear the $\alpha\beta$ T cell receptor complex ($\alpha\beta$ TCR) develop in the thymus from progenitors that originate in the hae-matopoeitic tissues (e.g. fetal liver, bone marrow, etc.). After colonisation, which in the mouse occurs from gestation day 11 onwards, these cells undergo a complex series of differentiation processes involving proliferation, TCR gene rearrangements, expression of accessory molecules and both pos-itive and negative selection of the TCR repertoire. These events lead to the generation of an MHC-restricted peripheral T cell pool and are thought to be regulated, at least in part, by the thymic stromal cells. Studies of the inter-actions between developing thymocytes and the stromal cells that make up the thymic microenvironment are therefore vital to our understanding of how T cell development is controlled *in vivo*. However, there is still consid-erable uncertainty about the role of individual stromal components and the investigation of specific stages in T cell development is hindered still further by the asynchronous manner in which maturation occurs *in vivo*.

To date, fetal thymic organ culture is the only system capable of support-ing a full programme of T cell development *in vitro*. Other systems that use isolated thymocytes support only limited parts of this programme (reviewed by Anderson & Jenkinson, 1995) and evidence exists to suggest that the way thymocytes perceive or respond to external stimulation may be modulated by the thymic microenvironment (Moore, Owen & Jenkinson, 1992). Attempts to investigate the interactions between thymocytes and stromal cells using either thymic stromal cell lines or primary monolayer cultures have also suffered from the fact that such preparations rarely maintain the *in vivo* characteristics of these cells. Indeed we have recently shown that mono-layer culture of MHC class II$^+$ thymic epithelial cells abrogates the ability of these cells to support positive selection and causes changes in gene expres-sion (Anderson *et al.*, 1998). In contrast, three-dimensional 'aggregate' culture of the same cells preserves their ability to support positive selection and prevents the changes in gene expression (Anderson *et al.*, 1998). Thus, the three-dimensional architecture of the thymus appears to be critical in promoting both homotypic and heterotypic interactions between the devel-oping thymocytes and the stromal cells, which are essential for the produc-tion of self-tolerant mature single positive (CD4$^+$ or CD8$^+$) thymocytes for exit into the periphery.

Conventional thymic organ cultures contain a complex heterogeneity of both stromal cells and immature lymphocytes. Detailed studies on the inter-actions between lymphocytes and individual thymic stromal cell populations necessitates a system where selected stromal cell types (e.g. epithelial cells and/or fibroblasts) can interact with purified populations of immature thymocytes under conditions where the normal cell-cell contacts of the *in vivo* thymus are maintained. Recently our group has developed a novel thymus organ culture system based on the enzymatic dissaggregation and reaggregation of embryonic thymus lobes and the immunomagnetic selection of defined populations of lymphoid and/or stromal cells. Purified thymic stromal cells such as MHC class II^+ cortical epithelial cells are prepared from embryonic thymus lobes rendered alymphoid by 2-deoxyguanosine (2-dGuo) and defined subpopulations of immature thymocytes are prepared by immunomagnetic selection. The mixed suspension is then spun down to form a pellet and the resuspended slurry placed into organ culture where reaggregation to form an intact lobe takes place within 12–18 hours (Jenkinson, Anderson & Owen, 1992). Importantly, these reaggregates support a complete programme of T cell development *in vitro* and allow individual stromal and lymphoid cells to interact in a three-dimensional manner comparable to the *in vivo* thymus.

General materials

70% ethanol/IMS for sterilising

Sterilised Nucleopore filters (0.8 µm, 13 mm diameter, Costar, Bucks, UK)

Tissue culture media (see below)

Dulbecco's phosphate buffered saline solution (Sigma, Poole, Dorset, UK)

Dulbecco's phosphate buffered saline solution without Ca^{2+} and Mg^{2+} (Sigma)

Sterilised and washed Artiwrap sponge foam (Medipost, Oldham, Lancs, UK)

Clear plastic sandwich boxes with flush-fitting lids (Stewart Plastics, Croydon, UK) with two small holes drilled in opposite ends of the lids

Dissection microscope with variable magnification – with both top lighting and trans-illumination

Dissection instruments, e.g. watchmaker's forceps, scissors and cataract knives (John Weiss, Milton Keynes, UK)

Gas cylinder: 10% CO_2 in air, set up to allow gassing of individual sandwich boxes with humidified gas (by passing through sterile distilled water)

Sterile tissue culture plastics, e.g. 45 mm and 90 mm Petri dishes, 10 ml pipettes, universals, 7 ml bijoux containers, 1 ml syringes and needles, etc.

Sterile lab plastics, e.g. pipette tips, Eppendorf tubes, etc.

2-mercaptoethanol (Sigma) made up as a stock of 7 μl in 20 ml PBS (i.e. 0.035%)

Cell culture quality 2-deoxyguanosine (Sigma) made up as a ~ 10 × stock in PBS (i.e. ~ 13.5 mM) and then filter-sterilised and frozen as suitable aliquots

Methods: preparation of tissue culture media and materials for organ culture

DMEM

Dulbecco's modified Eagles medium containing 4.5 g/l glucose, 110 mg/l sodium pyruvate, pyridoxine.HCl and $NaHCO_3$ but without glutamine (Sigma). Supplemented with the following (all given as final concentrations):

4 mM glutamine (from a 200 mM stock; Sigma)

100 U/ml penicillin and 0.1 mg/ml streptomycin (from a stock containing 5000 U/ml penicillin and 5 mg/ml streptomycin; Sigma)

10% v/v heat-inactivated FCS (Advanced Protein Products, Brierley Hill, West Midlands)

0.00035% 2-mercaptoethanol (from a 100 × stock in PBS; see above)

1% non-essential amino acids (from a 100× stock; Sigma)

10 mM HEPES buffer (from a 1 M stock; Sigma)

RF-10

RPMI 1640 medium containing 20 mM HEPES and L-glutamine, without $NaHCO_3$ (Sigma). Supplemented with the following (all given as final concentrations):

2 mM glutamine (from a 200 mM stock; Sigma)

100 U/ml penicillin and 0.1 mg/ml streptomycin (from a stock containing 5000 U/ml penicillin and 5 mg/ml streptomycin; Sigma)

10% v/v heat-inactivated FCS (Advanced Protein Products)

Nucleopore filters (0.8 μm):

These need to be sterilised before organ culture. Boil in a large volume of distilled water for at least 15 min and transfer (in a flow hood) to a sterile Petri dish. Allow to dry before using.

Artiwrap:

Cut the foam into small pieces just bigger than the Nucleopore filters and boil for at least 30 min in a large volume of distilled water. Change the water and boil the sponges for another 30 min and then change the water again and boil for a further 30 min. Transfer the sponges (in a flow hood) to a sterile Petri dish and allow to dry before using (this can take quite a long time, so they are best prepared the day before they are needed). These changes of water are essential to remove contaminants in the sponge that are otherwise cytotoxic.

Organ culture methods

Introduction

The essential feature of traditional murine fetal thymic organ culture is the maintenance of the explanted rudiment or fragment at the air–liquid interface, where exchange with both the medium below and the air above are optimal. Polycarbonate filters (0.8 μm pore size) provide an ideal support for explanted thymus lobes because they allow the movement of nutrients and macromolecules from below. These filters also provide a surface that discourages cell outgrowth, thereby maintaining the three-dimensional structure of the thymus, which is essential for normal thymocyte development. It is possible to float the filters directly on the surface of the medium, but they are best placed on foam supports to avoid the risk of them sinking, which impairs thymocyte development and causes necrosis. Gelatine foam sponges such as Sterispon have been widely used in the past, but synthetic foam sponges such as Artiwrap (1 mm thick) provide a satisfactory alternative so long as they are thoroughly washed before use (see above).

Most organ cultures use intact murine fetal thymic rudiments isolated at gestation days 13–15. At this stage, the individual lobes have recently been colonised by precursors derived from the haematopoeitic tissues that have the potential to give rise to both $\alpha\beta$ and $\gamma\delta$ T cell lineages as well as both dendritic cells and macrophages. During the 7-day organ culture period, a tightly regulated programme of events takes place including maturation, proliferation, TCR gene rearrangement and both positive and negative selection of the developing thymocytes. These developmental processes resemble those normally seen *in vivo*. Organ culture therefore provides a system in which it is possible to follow the progress of development (of both the lymphoid and stromal cell compartments) in the absence of further precursor recruitment and without the exit of the more mature thymocytes into the periphery. In addition, the small volumes of reagent that are required for fetal thymic organ cultures and the relative ease with which the tissue can be handled once it has been dissected from the embryo provides an accessible model with which to study the effects of a variety of treatments on early T cell development.

An adaptation of the organ culture method involves culturing gestation day 15 lobes for 5–7 days in the presence of 1.35 mM (2-dGuo). This reagent depletes the lobes of their lymphoid cells, providing 'empty' lobes for the introduction of selected lymphoid populations. Since the method was originally developed in our laboratory (Jenkinson et al., 1982), a number of other groups have introduced cells into 2-dGuo treated lobes by co-culture in hanging drops, formed by inverted Terasaki plates, followed, after a few hours, by transfer to normal organ culture. Lobes cultured in the presence of 2-dGuo are also used as a source of thymic stromal cells, since these cells are, at least in the short term, not affected by the 2-dGuo. Indeed we have recently shown that the proliferative capacity of thymic epithelial cells is not affected by a 5-day culture in the presence of 2-dGuo (Anderson et al., 1998).

Another adaptation of the organ culture method involves culturing the lobes under the surface of the media in so-called submersion-organ cultures (S-OC). Some time ago it was shown that T cell development occurred in S-OCs of murine fetal thymus lobes, but that the efficiency of cell growth was severely impaired compared to air–liquid interface cultures (Tomana et al., 1993). In addition, the T cells generated in these S-OCs were predominantly of the $\gamma\delta$ lineage, and as such this method has been little used. More recently, an adaptation of the original method was described, which involves culturing the lobes in S-OCs in an environment containing 60–80% O_2 (Watanabe & Katsura, 1993; Dou et al., 1995). Single fetal thymus lobes are submerged in 0.2 ml complete medium in 96-well U-bottomed plates and the plates placed in plastic bags and the air exchanged with a gas mixture of 5% CO_2, 70% O_2 and 25% N_2. T cell development in these high-oxygen submersion cultures is reported to be comparable to that in air–liquid interface cultures (Dou et al., 1995), but the traditional air–liquid interface organ culture remains the most popular way of culturing explanted murine thymus lobes to date.

One of the main problems with fetal thymus organ culture is that each individual thymus lobe provides only 0.5×10^6 lymphoid cells and $\sim 60\ 000$ stromal cells after the 7-day culture period. Many experiments therefore require large numbers of these lobes to obtain sufficient material for detailed analyses. With this in mind, we recently adapted the basic organ culture method to facilitate the in vitro culture of newborn thymus lobes, which contain up to $\sim 15 \times 10^6$ thymocytes. In this system, newborn lobes are removed and cleaned of adhering connective tissue by dissection. Each individual lobe is then bisected longitudinally with a single clean cut and the fragments transferred to organ culture. Bisecting the lobes is necessary to maintain lymphoid cell viability: without it the centre of the lobe becomes anoxic and, after 2 days, many of the thymocytes have undergone apoptosis. As many as 4 million lymphoid cells are provided by each newborn thymus fragment, and, since most of the culture periods are for no more than 48 h, a small culture dish containing 1.8 ml medium is sufficient to maintain cell viability. Small volumes of reagents are therefore sufficient for newborn thymus organ culture and, since the majority of newborn thymocytes have a $CD4^+CD8^+$ double positive phenotype, newborn organ cultures provide an ideal model in which to study the mechanisms that regulate thymocyte survival. Indeed, we have recently used this technique to investigate some of the biochemical mechanisms involved in the induction of thymocyte apoptosis by a variety of agents such as anti-CD3, anti-Fas and elevation of cAMP (Anderson *et al.*, 1996).

Although most of our studies have been carried out using pathogen-free BALB/c mice ($H-2^d$), murine thymic organ culture is not strain-specific and can be carried out with other strains of mice.

1. Murine fetal thymic organ culture (FTOC)

Isolation of thymic rudiments

1 Remove the uterus (gestation day 14–15) and place in a sterile container (e.g. a universal) for transport to a flow hood for further manipulation.

2 Remove the embryos from the uterus in their amniotic sacs and wash in sterile PBS (all subsequent procedures are carried out with the tissue immersed in either PBS or medium to prevent drying).

3 Remove the amniotic sacs using sterile watchmaker's forceps and transfer the embryos to a Petri dish containing RF-10.

4 Decapitate the embryo and position the body on its back with the limbs uppermost. In the 14–15 day embryo, the thymic rudiment is situated in the midline above the heart, on either side of the trachea (Fig. 3.1a).

5 Open the anterior surface of the chest using the forceps to reveal the thoracic cavity and remove the entire thoracic tree (heart, lungs, trachea and thymus) by grasping gently below the heart and place the tissue in a dish containing fresh RF-10 (see Fig. 3.1b).

6 Remove the two thymus lobes from the heart and lungs and clean off any excess connective tissue before transferring the lobes to organ culture.

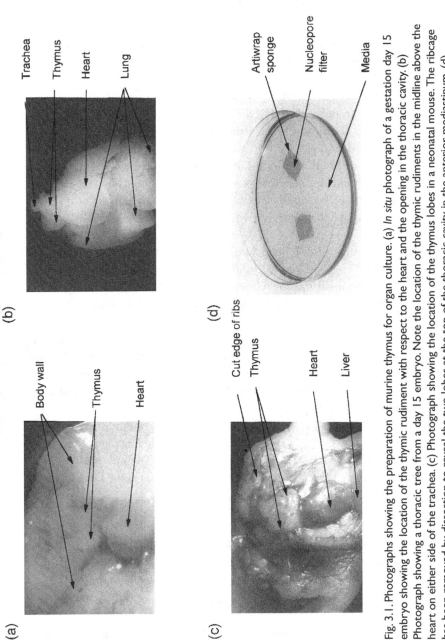

Fig. 3.1. Photographs showing the preparation of murine thymus for organ culture. (a) *In situ* photograph of a gestation day 15 embryo showing the location of the thymic rudiment with respect to the heart and the opening in the thoracic cavity. (b) Photograph showing a thoracic tree from a day 15 embryo. Note the location of the thymic rudiments in the midline above the heart on either side of the trachea. (c) Photograph showing the location of the thymus lobes in a neonatal mouse. The ribcage has been removed by dissection to reveal the two lobes at the top of the thoracic cavity in the anterior mediastinum. (d) Photograph showing the arrangement of filters and Artiwrap for FTOC. The arrangement is exactly the same for NTOC except that a 45 mm Petri dish is used with 1.8 ml medium and just one sponge and filter.

Organ culture of isolated thymic rudiments

1 Add 4.5 ml DMEM to a sterile 90 mm Petri dish and carefully add the required number of pieces of presterilised Artiwrap to the dish of medium. Normally three pieces of Artiwrap per Petri dish is the maximum for a 7-day culture period.

2 After a few seconds, turn the pieces of Artiwrap over and place an individual 0.8 μm polycarbonate Nucleopore filter on the top of each piece of Artiwrap (N.B. do not turn the filter over – simply place it on the Artiwrap and leave it).

3 Using either forceps, fine knives or a finely drawn mouth-controlled glass pipette, arrange up to six lobes on the surface of each Nucleopore filter (see Fig. 3.1d). Try not to 'squeeze' the lobes during transfer or this may cause damage to the tissue.

The following steps depend upon the type of incubator available. If humidified CO_2-gassed incubators are to be used, the Petri dishes can just be placed in the incubator (37 °C) and left for the desired time. If dry, CO_2-free incubators are used, steps 5–7 (below) need to be followed.

5 Transfer the Petri dishes to a clean sandwich box containing a platform and sterilised distilled water. Normally as many as three Petri dishes are placed in each sandwich box, such that each box contains as many as 54 individual thymus lobes. Seal around the lid with insulating tape, but leave the holes in the lid open.

6 Gas up the box for 10–15 min with 10% CO_2 in air (humidified by passing through sterile distilled water) to allow the medium to come to the appropriate pH and then seal over the holes with tape. Thymus organ cultures are particularly sensitive to pH, so it is important to gas the DMEM up to ~pH 7.2–7.4.

7 Transfer the boxes to the incubator (37 °C) and leave for the desired time.

In our experience, using dry, CO_2-free incubators gives the most satisfactory control over media pH and helps to limit the spread of infection

2. Newborn thymic organ culture (NTOC)

Newborn murine thymus lobes are much larger than fetal ones and as such are relatively easy to isolate from the surrounding tissue. A pair of fine knives

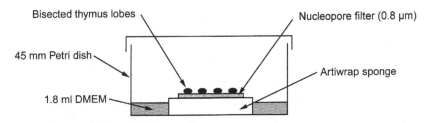

Fig. 3.2. Neonatal thymic organ cultures.

(such as cataract knives) is needed in addition to forceps and scissors because the lobes need to be cut in two before organ culture. A microscope that is illuminated from above is also much easier to use than one with trans-illumination because neonatal tissue is too dense to allow the passage of light.

Isolation and organ culture of bisected neonatal thymus lobes

1 Remove the thymus from the thoracic cavity with sterile scissors and a pair of fine forceps. The lobes are situated in the anterior mediastinum at the top of the thoracic cavity and have a characteristic opalescent appearance (see Fig. 3.1c). The lobes have a region of connective tissue between them, which can be grasped to remove the lobes. It is best not to grasp the lobes too tightly with the forceps or they may become damaged. Sometimes the two lobes will separate, in which case take out one at a time.
2 Transfer the thymus (or the individual lobes) to a Petri dish containing RF-10 and clear away any connective tissue.
3 With a single clean cut, carefully bisect each thymus lobe longitudinally. This technique is tricky and takes some practice! From this stage onwards the lobes have an open surface and are therefore very delicate. They are best handled with blunt knives rather than forceps because any pressure forces the cells out of the thymus.
4 Put 1.8 ml DMEM into a 45 mm Petri dish and place a piece of pre-sterilised Artiwrap into the medium. After a few seconds, turn the Artiwrap over and put a pre-sterilised filter on top of the sponge support (see Fig. 3.2).
5 Transfer as many as four pieces of thymus onto the surface of each filter. If possible lie the lobes on the filter so than the cut surface is uppermost.
6 Proceed with sealing and gassing the boxes, etc. exactly as described for fetal thymus organ cultures (see above).

3. Reaggregate thymus organ cultures

Introduction

Previous efforts to establish an *in vitro* system for studying the interactions between defined thymic stromal cells types and purified lymphocyte populations have been hampered by difficulties in maintaining the normal characteristics of thymic stromal cells in monolayer culture and by the fact that thymocytes in suspension culture respond quite differently to certain external stimuli than their counterparts maintained within the thymic microenvironment (Moore *et al.*, 1992). Reaggregate cultures rely on the enzymatic dissagregation of immature thymus lobes followed by the immunomagnetic selection of individual cell types, which are then reassociated to form a 'reaggregate' thymus lobe. Reformation of the intact 'reaggregate' lobe takes approximately 12–18 h.

Murine fetal thymus lobes (gestation day 15) depleted of lymphoid cells by culturing for 5–7 days in the presence of 2-dGuo are used as a source of thymic stromal cells for reaggregate cultures. These lobes are smaller than their lymphoid counterparts and often contain characteristic cystic structures. After trypsinisation, CD45 depletion removes any remaining macrophages and/or lymphocytes to give a suspension of thymic stromal cells. Individual cell types are then purified from this heterogeneous suspension by immunomagnetic selection (see Fig. 3.3). Purified populations of developing thymocytes are similarly prepared by immunomagnetic selection, but using a teased suspension of immature thymocytes from lobes of the appropriate gestational age (depending upon the thymocyte population required) as a starting population. Once the appropriate populations have been purified they are spun down together and the pellet transferred to organ culture as a slurry using a finely drawn mouth-controlled glass pipette. Reassociation to form a 'solid' reaggregate thymus organ then occurs overnight. This approach has been used to show that the initial development of CD4$^-$8$^-$ double negative thymocytes requires the support of both thymic epithelium and mesenchyme cells (fibroblasts) to reach the CD4$^+$8$^+$ double positive stage, but that thereafter MHC class II$^+$ cortical epithelial cells are sufficient for the further development of DP cells (Anderson *et al.*, 1993). More recently we have used the same technique to show that down-regulation of CD44 marks the last stage of thymocyte development to be dependent upon fibroblast support (Anderson *et al.*, 1997). The same technique has also been used to define more accurately the events that occur during the maturation of CD4$^+$CD8$^+$TCR$^-$ thymocytes (Anderson *et al.*,

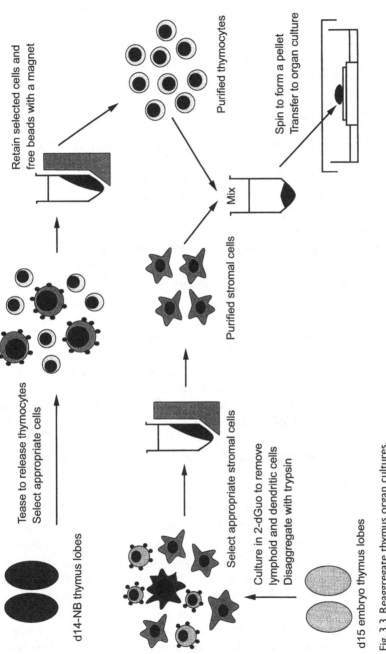

Tease to release thymocytes
Select appropriate cells

d14-NB thymus lobes

Retain selected cells and
free beads with a magnet

Purified thymocytes

Select appropriate stromal cells

Purified stromal cells

Mix

Culture in 2-dGuo to remove
lymphoid and dendritic cells
Disaggregate with trypsin

d15 embryo thymus lobes

Spin to form a pellet
Transfer to organ culture

Fig. 3.3. Reaggregate thymus organ cultures.

1994). The method for purifying $CD4^+CD8^+TCR^-$ thymocytes and MHC class II^+ thymic epithelial cells, together with the technique for preparing reaggregate thymus lobes is described here. Because thymocytes are susceptible to apoptosis it is important that all centrifugations and cell purifications are carried out at $4°C$.

Specific reagents

Dynal beads (sheep anti-rat IgG beads and goat anti-mouse IgG beads; Dynal, Wirral, UK)

Magnetic particle collector (for Eppendorfs, Dynal)

Trypsin ($10 \times$ solution (i.e. 2.5%) aliquoted and frozen)

Pronase E (Cell Culture Reagent, Sigma)

Detachabead (Dynal)

Soft glass tubing for micropipettes (product code TWL-290-041G: Fisons, Loughborough, Leics, UK)

Aspirator tube assembly for microcapillary pipettes (Sigma)

Freezing vials

Antibodies for bead coating:

Anti-I-A^d (clone MK-D6; Becton Dickinson, Cowley, Oxford, UK)

Pan anti-CD45 (clone M1/9; ATCC)

Anti-CD8 (clone YTS169.4; Harlan Seralabs, Crawley Down, UK)

Anti-CD3 (clone KT3; Serotec, Kidlington, Oxford, UK)

Anti-medullary epithelium (clone A2B5; a gift from Prof. M. Raff, University College, London, UK)

Immunomagnetic separation of stromal and lymphoid cells

Introduction

Isolation of stromal and lymphoid cells is achieved using antibody-coated magnetic beads. These are supplied pre-coated with an anti-species antibody (e.g. sheep anti-rat) and are then coated with an appropriate second antibody (e.g. rat anti-mouse CD45). This gives beads that are effectively coated with the secondary antibody (in this case anti-mouse CD45). For immuno-magnetic selection, cells are resuspended in 200 µl RF-10 and transferred to a round-bottomed freezing vial to give a greater surface area over which the cells and beads can interact. For negative selection, i.e. depletion of unwanted cells, beads are added to give a bead:cell ratio of 10:1 to ensure maximum rosetting. To promote interactions, cells and beads are spun together for 10 min (1000 rpm, $4°C$). This results in the formation of bead/cell aggregates, or rosettes. Rosetted cells and free beads are then removed using a magnetic particle collector (Dynal) and the free cells collected and washed (see Fig. 3.3). Cells are commonly centrifuged with multiple rounds of antibody-coated beads to achieve maximal depletion.

Following depletion of unwanted cells, the remaining cell suspension is then commonly used in the positive selection of the desired cell population. Bead:cell interactions are performed as described above except that the Dynal beads are added to give a bead:cell ratio of approximately 3:1. Rosetted cells are then collected by magnet and washed a couple of times to ensure that no carry-over of non-specifically bound cells occurs. Beads are then removed using trypsin, Pronase or Detachabead.

Bead coating procedure

1 Take up the required volume of beads in 1 ml RF-10 HEPES (in a 1.5 ml Eppendorf tube) and mix by gentle pipetting.
2 Transfer the tube to the magnet and collect the beads to one side of the tube. Remove the media and resuspend the cells in 1 ml fresh RF-10.
3 Repeat this washing procedure 3 times to ensure that the beads are free of azide, which is cytotoxic.
4 Resuspend the beads in the appropriate antibody solution and leave to incubate at 4 °C overnight.
5 Wash the beads in 3 × 1 ml RF-10 and resuspend in the starting volume of medium (i.e. if you start with 100 μl of beads, they should be resuspended in 100 μl RF-10 immediately before use).

Examples of specific bead coating procedures are given in Tables 3.1 and 3.2.

Preparation of CD45-depleted thymic stromal cells from 2-dGuo-treated thymus lobes

1 Remove the thymus lobes from organ culture and transfer them to a sterile Eppendorf in Ca^{2+}/Mg^{2+}-free PBS. Allow the lobes to settle to the bottom and carefully remove the PBS with a pipette.
2 Repeat this procedure 3 times and resuspend the lobes in 600 μl 1:10 diluted trypsin (in EDTA). Incubate for 25–30 min at 37 °C.
3 Dissaggregate the lobes by gently pipetting them up and down with a 1 ml pipette tip. Take care to be very gentle at this stage because cells in trypsin are very delicate. If the lobes do not disaggregate completely, put the tube back in the incubator for another 5 min and repeat the pipetting.

Table 3.1. *Bead coating protocols for isolating stromal cells from 2-dGuo treated thymus lobes*

Use	Procedure
Depletion of haematopoeitic cells from thymic stromal cells	100 μl of anti-rat IgG beads in 1 ml anti-CD45 culture supernatant (clone M1/9)
Depletion of MHC class II$^+$ thymic epithelium (Class II haplotype is species specific: I-A^d in Balb/c)	300 μl anti-mouse IgG beads in 1.5 ml 1:50 diluted anti-I-A^d (clone MK-D6)
Depletion of A2B5$^+$ medullary epithelium	100 μl anti-mouse IgG beads in 1 ml A2B5 culture supernatant
Positive selection of MHC class II$^+$ thymic epithelium	100 μl anti-mouse IgG beads in 500 μl of 1:500 diluted anti-I-A^d (clone MK-D6)

4 Once a suspension of cells has been obtained, inactivate the trypsin by adding 1 ml RF-10 to the tube and mix gently.

5 Spin the cells down (1000 rpm, 10 min), resuspend in 200 μl RF-10 and transfer to a freezing vial.

6 Add 100 μl anti-CD45 coated Dynal beads and spin for 10 min at 1000 rpm. Gently resuspend the pellet by pipetting and repeat the centrifugation and pipetting twice.

7 Resuspend the cells, transfer to an Eppendorf and remove the rosetted cells with the magnet. Wash the rosettes and free beads with 0.5 ml RF-10 and again recover the un-rosetted cells with the magnet (this recovers any unrosetted cells trapped in the pellet).

8 Spin the cells down, resuspend in 1 ml fresh RF10 and count.

Preparation of MHC class II$^+$ thymic cortical epithelial cells from CD45-depleted thymic stromal cells

1 Prepare CD45-depleted thymic stromal cells as described above, resuspend in 200 μl RF-10 and transfer to a freezing vial.

2 Add 100 μl A2B5 coated anti-mouse IgG beads and spin three times, resuspending the cells and beads in between each spin.

3 Remove the rosetted cells on the magnet and retain the supernatant. Wash the beads with 100 μl RF-10 and pool the two supernatants.

4 Count the remaining cells and add the appropriate volume of anti-IA^d coated Dynal beads and spin again for 10 min.

Table 3.2. *Bead coating for the preparation of CD4⁺CD8⁺ thymocytes from newborn thymus*

Use	Procedure
Depletion of CD3⁺ thymocytes	300 μl anti-rat IgG beads in 1 ml anti-CD3 culture supernatant (clone KT3)
Positive selection of CD8⁺ cells	100 μl anti-rat IgG beads in 500 μl 1:100 diluted anti-CD8 (clone YTS169.4)

5 Resuspend the cells and beads and then spin again. Collect the rosetted cells by magnet and discard the free cells.
6 Wash the rosetted cells three times in 600 μl PBS and resuspend in 50 μl PBS. Add 150 μl of 10 mg/ml ice-cold pronase and incubate for 2 min in a water bath at 37°C, pipetting gently after 1 min.
7 Inactivate the pronase by adding 800 μl ice-cold RF-10 and remove the free beads and remaining rosetted cells on the magnet.
8 Collect the liberated cells by centrifugation and count.

Preparation of CD4⁺CD8⁺ thymocytes from newborn thymocytes

1 Prepare a suspension of newborn thymocytes by gently teasing newborn thymus lobes. Count the cells and retain 1.2×10^7 cells (the thymus from one newborn mouse contains up to $\sim 1.5 \times 10^7$ cells). Spin the cells down and resuspend in 200 μl RF-10.
2 Transfer the suspension to a freezing vial and add 100 μl of anti-CD3 coated beads. Spin the cells and beads down and remove the rosetted cells on a magnet. Because so many of the cells are CD3⁺ there is no need to spin the cells and beads down three times on either the first or second rounds of depletion since most of the beads will become rosetted within one or two spins.
3 Retain the free cells and repeat the depletion using 100 μl anti-CD3 coated beads, but this time spin the cells and beads together twice before removing the rosetted cells on the magnet.
4 Retain the free cells and repeat the depletion procedure once more using 100 μl beads and three spins before removing the rosetted cells on the magnet. If at any stage the volume of free cells is over 300 μl, spin the cells down without the beads and take them up in 200 μl medium to continue: if the volume of cells and beads reaches more than ~ 400 μl then the efficiency of depletion is reduced.

5 Count the recovered cells, spin and take up in 200 µl RF-10 and transfer to another freezing vial.
6 Add the appropriate volume of anti-CD8 coated beads to give a final ratio (beads:cells) of 3:1. Spin the cells and beads together, resuspend the pellet and spin again. Collect the rosetted cells by magnet and wash three times in 600 µl Ca^{2+}/Mg^{2+}-free PBS.
7 Resuspend the rosetted cells in 600 µl 1:10 diluted trypsin (in EDTA) for 2 min at 37°C, pipetting gently after 1 min.
8 Check under the microscope that the beads have come off the cells and stop the trypsin with 1 ml RF-10 media. Remove the beads and remaining rosettes by magnet.
9 Spin the cells down, take up in 1 ml RF-10 and count. The expected recovery is around 20–40% (i.e. $2.4–4.8 \times 10^6$ cells).

Formation of reaggregate thymus organ cultures

Reaggregates are formed from freshly isolated stromal and lymphoid preparations mixed at the desired ratio, usually ~1:1–1:2. If fibroblasts are also added, the ratio of fibroblasts: epithelial cells: T cell precursors is ~1:2:2. Cultures should contain no more than 3×10^6 cells in total to prevent the centre of the lobe becoming anoxic, so it is better to set up two small reaggregates than one big one if larger numbers of cells are required.

1 Set up the required number of Petri dishes containing 4.5 ml DMEM and Artiwrap sponges overlaid by nucleopore filters exactly as described for organ cultures.
2 Mix the desired number of lymphocytes and stromal cells in 1 ml RF-10 and spin the cells down to form a pellet.
3 Remove as much of the media as is possible (without disturbing the pellet) and then vortex the pellet to form a slurry.
4 Draw the slurry up a fine mouth-controlled glass pipette and carefully expel the suspension onto the surface of a filter to form a discrete standing drop. Control the rate of expulsion so that any excess fluid has time to pass through the filter so that the drop does not spread out: this is crucial to the successful reaggregation. No more than one reaggregate should be placed on each filter.
5 Transfer the Petri dish into a sandwich box and gas up exactly as described for the standard organ cultures. Intact lobes reform from these mixtures within 12–18 h. In the case of reaggregates made with CD4$^+$8$^+$CD3$^-$ thymocytes, evidence of positive selection is best observed after 3–4 days.

4. Analysis of thymocyte phenotype by flow cytometry

Specific reagents

Anti-CD8 FITC conjugate (clone 53-6.7, PharMingen, San Diego, CA, USA)
Anti-CD4 PE conjugate (clone RM4-5, PharMingen)
Anti $\alpha\beta$-TCR (clone H57.597, PharMingen)
Streptavidin-APC (PharMingen)
Anti-hamster Fab$_2$ biotin-conjugate (cross-absorbed against rat/mouse; Caltag, San Francisco, CA, USA)
1% paraformaldehyde in PBS
FACS tubes

Methods

1 Put ～1 ml RF10 on the lid of a sterile Petri dish and collect the lobes/reaggregates from organ culture into this medium.

2 Using fine knives, gently tease the lobes to release the thymocytes and then transfer the suspension to a sterile Eppendorf tube. Take care not to include the material that does not tease up since this is mainly stromal cells. Wash the lid with a further 0.5 ml medium and gently pipette the suspension up and down the pipette tip to disperse any small clumps.

3 Stand the suspension on ice for ～5 min to allow any remaining clumps to settle (this is important as they can block the tubes inside the flow cytometer). Remove the suspension into a clean Eppendorf, taking care not to pick up any of the clumps and then count the cell density using a haemocytometer slide. Correcting for any loss of medium along the way, calculate the number of cells recovered from each lobe/reaggregate. Stand the suspension on ice.

4 Tease a newborn lobe and retain approximately 1 million cells (also cleared of any clumps) in a second Eppendorf tube. These cells will serve as a control for calibrating the flow cytometer.

5 Spin the cells down (1000 rpm, 10 min) and resuspend the pellet in ～50 μl anti $\alpha\beta$-TCR (diluted as appropriate). Stand the cells on ice for 30 min and then top up the tube with 1 ml ice-cold PBS.

6 Spin the cells down and resuspend the pellet in 50 μl anti-hamster biotin. Stand the cells on ice for 30 min.

7 Wash the cells in 1 ml PBS and resuspend them in 50 μl of a mixture of anti-CD4 PE, anti-CD8 FITC and streptavidin-APC (all diluted as appropriate). Stand the cells on ice for 30 min (these reagents are light-sensitive, so they must be kept in the dark.)

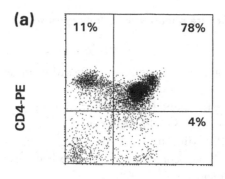

CD8-FITC

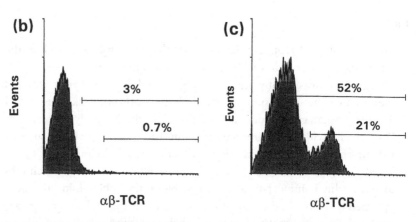

Fig. 3.4. A typical CD4/CD8 profile of newborn mouse thymocytes and a typical $\alpha\beta$-TCR profile of adult mouse thymocytes. (a) Newborn mouse thymus lobes were teased apart with fine knives and the released thymocytes stained with a mixture of PE-conjugated anti-CD4 and FITC-conjugated anti-CD8 for 30 min on ice. Cells were then fixed in 1% paraformaldehyde (in PBS) and analysed by flow cytometry. (b), (c) Thymocyte suspensions were prepared from adult mouse thymus lobes and stained with anti-$\alpha\beta$ TCR for 30 min on ice. Cells were then washed in PBS and incubated for 30 min (on ice) with streptavidin-FITC. Cells were then washed in PBS, fixed in 1% paraformaldehyde and analysed by flow cytometry. Profile (b) shows the background staining obtained from cells incubated with just the secondary antibody (streptavidin-FITC) while profile (c) shows the positive staining obtained from cells incubated with both the primary (anti-$\alpha\beta$) and secondary antibodies. Sample (b) was used to set up the flow cytometer and to show the cut-off between the negative and positive populations. Panel (c) shows that there are three distinct populations of $\alpha\beta$-thymocytes in adult mice: $\alpha\beta$-negative, $\alpha\beta$-low ($\sim$31%) and $\alpha\beta$-high ($\sim$21%). Background levels of staining must be subtracted from the percentage depicted in panel (c) to obtain the true percentages given above.

8 Wash the cells in 1 ml PBS and resuspend the pellets in a small volume of PBS (~50 μl). Add 200 μl 1% paraformaldehyde and mix thoroughly. Transfer the suspensions to FACS tubes and wrap each one in foil. Keep the samples in the fridge until they can be analysed. Although the samples will keep for a few days in the fridge, they are best analysed as soon as is convenient. Figure 3.4 shows typical CD4 and CD8 profile of a newborn thymus and a typical adult thymocyte $\alpha\beta$ TCR profile.

9 Once the cells have been analysed by FACS, correct the percentage of cells in each gate to absolute cell numbers. Some reagents do not affect the overall percentage of $CD4^+$ and $CD8^+$ cells in each culture, but cause drastic effects on the overall survival, so it is important to calculate the absolute number of each cell phenotype recovered from the culture.

References

Anderson, G. & Jenkinson, E. J. (1995). The role of the thymus during T-lymphocyte development *in vitro*. *Semin. Immunol.*, **7**, 177–83.

Anderson, G., Jenkinson, E. J., Moore, N. C. & Owen, J. J. T. (1993). MHC class II positive epithelium and mesenchyme cells are both required for T cell development in the thymus. *Nature*, **362**, 70–3.

Anderson, G., Owen, J. J. T., Moore, N. C. & Jenkinson, E. J. (1994). Characteristics of an *in vitro* system of thymocyte positive selection. *J. Immunol.*, **153**, 1915–20.

Anderson, K. L., Anderson, G., Michell, R. H., Jenkinson, E. J. & Owen, J. J. T. (1996). Intracellular signaling pathways involved in the induction of apoptosis in immature thymic T-lymphocytes. *J. Immunol.*, **156**, 4083–91.

Anderson, G., Anderson, K. L., Tchilian, E. Z., Owen, J. J. T. & Jenkinson, E. J. (1997). Fibroblast dependency during early thymocyte development maps to the $CD25^+44^+$ stage and involves interactions with fibroblast matrix molecules. *Eur. J. Immunol.*, **27**, 1200–6.

Anderson, K. L., Moore, N. C., McLoughlin, D. E. J., Jenkinson, E. J. & Owen, J. J. T. (1998). Studies on thymic epithelial cells *in vitro*. *Dev. Comp. Immunol.*, **22**, 367–77.

Ashton-Rickardt, P. G., Van Kaer, L., Schumacher, T, N, M., Ploegh, H. L. & Tonegawa, S. (1993). Peptide contributes to the specificity of positive selection of $CD8^+$ T cells in the thymus. *Cell*, **73**, 1041–9.

Crompton, T., Gilmour, K. C. & Owen M. J. (1996). The MAP kinase pathway controls differentiation from double negative to double positive thymocyte. *Cell*, **86**, 243–51.

Dou, Y.-M., Watanabe, Y., Aiba, Y., Wada, K. & Katsura, Y. (1995). A novel culture system for induction of T cell development: modification of fetal thymus organ culture. *Thymus*, **23**, 195–207.

Hogquist, K. A., Gavin, M. A. & Bevan, M. J. (1993). Positive selection of CD8 +

T cells induced by major histocompatibility complex binding peptides in fetal thymic organ cultures. *J. Exp. Med.*, **177**, 1469–73.

Jenkinson, E. J. & Anderson, G. (1994). Fetal thymic organ cultures. *Curr. Opin. Immunol.*, **6**, 293–7.

Jenkinson, E. J., Franchi, L. L., Kingston, R. & Owen, J. J T. (1982). Effect of deoxyguanosine on lymphopoiesis in the developing thymus rudiment *in vitro*: application in the production of chimeric thymus rudiments. *Eur. J. Immunol.*, **12**, 583–7.

Jenkinson, E. J., Anderson, G. & Owen, J. J. T. (1992). Studies on T cell maturation on defined thymic stromal cell populations *in vitro*. *J. Exp. Med.*, **176**, 845–53.

Levelt, C. N., Ehrfeld, A. & Eichmann, K. (1993a). Regulation of thymocyte development through CD3. 1. Timepoint of ligation of CD3ϵ determines clonal deletion or induction of developmental program. *J. Exp. Med.*, **177**, 707–16.

Levelt, C. N., Mombaerts, P., Iglesias, A., Tonegawa, S. & Eichmann, K (1993b). Restoration of early thymocyte differentiation in T cell receptor β chain deficient mutant mice by transmembrane signalling through CD3ϵ. *Proc. Natl. Acad. Sci. USA*, **90**, 11401–5.

Moore, N. C., Owen, J. J. T. & Jenkinson, E. J. (1992). Effects of the thymic microenvironment on the response of thymocytes to stimulation. *Eur. J. Immunol.*, **22**, 2533–7.

Merkenschlager, M. & Fischer, A. G. (1994). In vitro construction of graded thymus chimeras. *J. Immunol. Meth.*, **171**, 177–88.

Tomana, M., Ideyama, S., Iwai, K., Gyotoku, J. I, Germeraad, W. T. V., Muramatsu, S. & Katsura, Y. (1993). Involvement of IL-7 in the development of $\gamma\delta$ T cells in the thymus. *Thymus*, **21**, 141–57.

Watanabe, Y. & Katsura, Y. (1993). Development of T cell-receptor $\alpha\beta$-bearing T cells in the submersion organ culture of murine fetal thymus at high oxygen concentration. *Eur. J. Immunol.*, **23**, 200–5.

4

T cells

Hergen Spits and Colin G. Brooks

Introduction

T cells are central to a functional immune system. This is dramatically illus-
trated by the greatly increased susceptibility to infectious diseases seen in
congenitally athymic mice, neonatally thymectomised mice, and in patients
who have a congenital lack of T cells. It is therefore not surprising that since
their discovery, T cells have been the subject of intense research. The development of methods to grow T cells *in vitro* in the mid-seventies was a major
breakthrough in the study of these cells. Over the past two decades *in vitro*
cultured lines of murine T cells have provided important insights into our
understanding of the role of T cells in immune responses. Since in the human
in vivo studies are difficult for obvious reasons, the capability of growing T
cells *in vitro* is even more critical. Among the significant contributions of
studies using lines of human CTL and Th cells are the generation of the first
anti-TCR antibodies (Meuer *et al.*, 1983). Moreover, T cell clones were used
to unravel the complexity of HLA class I and class II polymorphism (see for
example, Spits *et al.*, 1982). A more recent hallmark is the discovery of
tumour-associated antigens like the melanoma-associated MAGE antigens
using human melanoma-specific CTL clones (van Pel *et al.*, 1995).

In this chapter we will discuss the general strategies employed for the
culture of T cell lines and clones, and provide details of some of the methods
used in our laboratories.

Growth factors

The identification of a single protein, IL-2, as the principal active T cell
growth factor (TCGF), coupled with the comparatively low cost of recombinant IL-2, has led to its widespread use for generating T cell lines. However,
it is now recognised that many other proteins have T cell growth-promoting

activity, including IL-4, IL-7, IL-9, IL-12 and IL-15, and several cytokines such as IL-1 and IL-6, which themselves lack TCGF activity, can nonetheless enhance the growth-promoting activity of IL-2 and other TCGFs. This implies the existence of a high degree of complexity in the natural regulation of T cell growth and differentiation. It follows that the use of any single growth factor may be suboptimal and that the choice of growth factor may have a substantial influence on the phenotype and functional potential of the resulting cells. It should not be assumed that different types of T cell have the same growth factor requirements. For example, some human CD4⁻CD8⁻ $\alpha\beta$ T cells can use IL-3 as a growth factor. Despite this, because of limitations of space, we shall concentrate in this chapter on systems and methods in which IL-2 is used as the exogenous growth factor. Recombinant human IL-2 is equally effective on both human and mouse T cells. The dose used is important: it should be sufficient to support the rapid expansion of T cells during the proliferation phase but not so high as to interfere with the entry of cells into the resting phase (see below). Typically, IL-2 would be used at 100 IU/ml during the expansion phase, and 10–20 IU/ml during the resting phase. It is advisable to determine the units needed by bioassay on the CTLL2 line in comparison with the WHO international IL-2 standard (National Institute of Biological Standards, South Mimms, UK) rather than relying on manufacturer's unitage values.

Establishment of T cell lines

The principle of culturing T cells *in vitro* is simple. Freshly isolated T cells are stimulated in a way that upregulates receptors for growth factors (Cantrell & Smith, 1984). Endogenous growth factors produced by the T cells themselves, or exogenous growth factors provided by the investigator at an appropriate concentration, allow the T cells to proliferate. It is important to recognise that one aspect of the physiology of T cells is that the induction of proliferation is followed by a spontaneous downregulation of the growth factor receptors, cessation of cell division, and an entry of the cells into a quiescent resting state, perhaps equivalent to the memory phase *in vivo*. To induce further rounds of proliferation, restimulation is required and long-term T cell lines and clones are best grown by subjecting them to a strict cyclic regimen involving successive stages of stimulation, proliferation, and imposed rest. In most cases where the functional capacity (proliferation, cytotoxicity, cytokine secretion) of T cell lines or clones is being studied or made use of, fully rested (small, non-proliferating) cells should be used in assays since at this stage of the maintenance cycle both antigen specificity and functional capabilities are maximal, and the likelihood of residual feeder cells being present is minimal. It is possible, at least in the mouse, to induce T cells, especially CD8 and $\gamma\delta$ T cells, to grow continuously in high concentrations of IL2. However, this invariably leads to loss of specificity and function, acquisition of unusual morphological features, and frequently also an eventual loss of growth capacity itself.

If one wishes to grow T cells without being interested in their specificity, one can stimulate them with a mitogen (in the mouse concanavalin A (ConA) gives a stronger and more general response, in the human phytohaemagglutinin (PHA) works better) or with a monoclonal antibody against the TCR/CD3 complex. Because the only cells in suspensions prepared from lymphoid organs or blood that will proliferate to any substantial extent under these conditions are T cells, there is actually no need to purify the T cells prior to culture unless the properties of T cells present in short-term cultures are being studied. Indeed, highly purified populations of T cells will hardly proliferate due to a requirement for factors that are provided by non-T cells, in particular by monocytes/macrophages and dendritic cells. These factors may include IL-1, IL-6, TNFα and IL-12, but other less well defined factors may be required as well. The need to provide these factors is one of the principal problems and major limitations in the long-term culture of T cells, and is generally achieved by the routine addition of appropriate preparations of other cells, usually referred to as accessory, feeder, or filler cells. Apart from the inconvenience, the problems associated with the use of feeder cells include (i) the need to prevent the feeder cells themselves proliferating, (ii) the potential for contamination of responder T cells with T cells from the feeder population and (iii) their possible influence on the nature and functional capabilities of the responder T cell populations generated.

The most commonly used feeder cells are, in the mouse, freshly prepared spleen cells, and, in the human, freshly prepared blood mononuclear cells. These must be rigorously depleted of T cells and/or treated either with doses of irradiation or with mitosis-inhibiting agents such as mitomycin C. For simplicity, and because factors produced by T cells themselves can be important contributors to the overall feeder effect, the latter procedure is almost invariably used. However, it must be borne in mind that irradiation and drugs not only inhibit proliferation but may also inhibit 'feeder effects'. Hence, the dose of irradiation or drug used should be chosen carefully, and what is optimal in one situation may not be optimal in another. In general, the lowest dose of irradiation or drug that prevents any detectable contamination of the T cell lines or clones with feeder-derived T cells should be used, typically 10–20 Gy (1000–2000 rad) of X-ray or gamma-ray irradiation. Various other types of cell have been used as feeders, such as mouse tumour cell lines, EBV-transformed human B cell lines, and dendritic cells. With the former two cell types, there is no problem of T cell contamination, but the powerful prolife-rative capacity of these cells usually necessitates radiation doses of 25–50 Gy. The dendritic cells cells, if adequately pure, would theoretically not require any anti-proliferative treatment.

As already noted, there is no need to purify T cells since the process of establishing a T cell line is itself a highly effective method of purifying T cells. However, it is also a highly selective procedure. Even when polyclonal acti-vators such as ConA, PHA, or anti-TCR/CD3 mAbs are used, a substantial bias towards either CD4 or CD8 T cells rapidly develops in cultures and often goes to completion. If lines of only one or other of these cell types are required, or if lines of minor populations of T cells, such as $\gamma\delta$ or CD4$^-$CD8$^-$ $\alpha\beta$ T cells, are sought, it is essential to purify the relevant pop-ulation before the establishment of cultures and/or at various times there-after. A large selection of purification procedures are available, but a description of these is beyond the scope of this chapter. The only exceptions to the statement that the only cells that will grow under the conditions used for the establishment of T cell lines are T cells themselves, are natural killer (NK) cells. Contamination by these cells can be quite substantial during the early stages of establishment of lines of $\gamma\delta$ and CD4$^-$CD8$^-$ $\alpha\beta$ T cells. However, because optimal growth of NK cells requires much higher con-centrations of IL-2 than are needed for the growth of T cells (see Chapter 7), and because they have a more limited growth potential than T cells, their proportion declines rapidly with time. If necessary, these cells should be removed prior to the start of culture. In the mouse a simple procedure is to use the NK-specific monoclonal antibody (mAb) 3A4 (Pharmingen, San Diego, CA, USA) and complement. In the human, depletion of NK cells can be achieved with anti-CD56 and magnetic beads (Dynal, Oslo, Norway).

If one wishes to culture antigen-specific T cells, freshly isolated T cells need to be stimulated with the relevant antigen. In the case of T cells specific for foreign major histocompatibility (MHC) antigens, the frequency of antigen-specific T cells in normal lymphocyte preparations is very high, of the order of 1/10–1/100. Alloantigen-specific cell lines can be readily gen-erated by using as feeder cells either allogeneic lymphocytes or EBV-transformed B cell lines expressing the appropriate antigens. By starting with CD4-depleted or CD8-depleted responder cells, one can generate CD8$^+$ class I-specific or CD4$^+$ class II-specific lines respectively. These will gener-ally be reactive, at least in the early stages, with all non-self donor class I or class II molecules. Lines specific for individual class I molecules can be generated by appropriate selection of responder/feeder cell combinations, or at a later stage by cloning. In the former case it should be noted that, although the line may be specific for the class I or class II molecule of interest, it will in fact be composed of T cells of many different specificities related to the different peptides present in the grooves of individual MHC molecules.

While it is easy to obtain allo-specific $CD4^+$ and $CD8^+$ T cell lines, establishment of T cell lines with specificity for individual proteins or peptides is less straightforward. The most likely reason for this is the low frequency of specific T cells. In general the frequency of T cells with specificity for a particular peptide–MHC complex is very low, of the order of 1/10000 or (much) lower. However, in immunised mice, recently vaccinated humans, or in individuals suffering from an infectious disease or cancer, frequencies can be as high as 1/1000. These antigen-specific T cells can often be further enriched by repetitive stimulation with antigen *in vitro*.

To generate $CD4^+$ antigen-specific T cell lines one simply adds the relevant antigen to cultures containing lymphocytes from an immune individual, followed by subsequent cycles of stimulation in which resting responder cells are restimulated with antigen in the presence of syngeneic or autologous feeder cells bearing the appropriate MHC molecules. In this situation the feeder cells not only provide a 'feeder effect' but also act, in the case of peptide antigens, as antigen-presenting cells, or in the case of protein antigens as both antigen-processing and antigen-presenting cells. In the human, the frequent severe limitation in the availability of autologous cells for use as feeders has led to the use of alternate strategies. Feeder cells bearing appropriate MHC molecules may be available from another donor. Minimal numbers of autologous antigen-presenting cells (APC), sufficient to provide a stimulus, may be mixed with an excess of unrelated feeder cells. In both these cases, at least during the early cycles of culture, there is the danger of inducing the proliferation of alloreactive T cells in addition to antigen-specific T cells. A frequently used procedure for the generation and maintenance of human T cell lines and clones is to establish either before or at the time of responder cell donation an EBV-transformed lymphocyte line from the donor for use as APC/feeder cell. However, as noted above, such lines require very high levels of radiation to destroy their proliferative activity, which probably compromises their innate antigen-processing and feeder cell capabilities, and seem to be particularly poor at presenting certain antigens, for example the 65 kDa hsp from *Mycobacterium leprae* (Ottenhof *et al.*, 1986). In addition, such lines have a tendency to stimulate T cell proliferation in an antigen–non-specific manner.

The generation of antigen-specific CD8$^+$ T cell lines presents additional problems due to the requirement for the relevant antigenic proteins to be processed via the class I pathway. This substantially restricts the range of antigens to which CD8$^+$ T cell lines and clones can be generated, in large part to antigens that are the natural components of, or which can be introduced by genetic manipulation into, viruses that infect feeder lymphomyeloid cells, for example influenza (Gotch, McMichael & Rothbarth, 1988) or HIV (Koenig et al., 1988). More recently strategies have been developed for inducing the expansion of tumour-specific CD8$^+$ T cells from tumour-bearing animals or from cancer patients by stimulation with irradiated tumour cells either directly or in admixture with autologous PBMC or EBV-transformed B cells (for example, melanoma-specific CTL (Van Pel et al., 1995). In cases where the precise identity of an antigenic epitope is known, CD8$^+$ T cell lines can be generated in a similar manner to CD4$^+$ lines by the direct addition of the peptide to cultures, where it binds to class I molecules without having to enter the class I antigen-processing pathway. Several protocols for generating peptide-specific CTL have been described recently. The methods are very similar but differ in details with respect to the use of medium or of growth factors: IL-2, IL-4 or IL-7 or combinations of these factors. There are also differences in the type of APC used by different authors. Some use total PBMC (Plebanski et al., 1995), others activated B cells (Celis et al., 1994) or monocytes that have been differentiated to dendritic cells with GM–CSF and IL-4 (Bakker et al., 1995) or EBV-transformed B cells (De Waal Malefyt et al., 1993). As none of these papers offer careful comparisons of different protocols it is unclear whether there are particular advantages of using one or another growth factor or APC type.

Finally, there has recently been the development of methods that permit the generation of antigen-specific T cell lines from normal unimmunised donors. The scarcity of antigen-specific T cells amongst lymphocytes from normal donors, and the enormous degree of enrichment required, demand relatively large-scale initial cultures and highly efficient antigen stimulations. This has been most commonly achieved either by using highly proficient APC, in particular dendritic cells, or APC that express exceptionally high levels of the relevant peptide–MHC complex, such as TAP-deficient cell lines exposed to high concentrations of specific peptide.

Generation of T cell clones

The ultimate aim of the T cell biologist is often the derivation of T cell lines derived from a single progenitor, i.e. T cell clones. Cloning represents the ultimate form of cell purification, although in the case of T cells the mandatory requirement for the periodic addition of feeder cells potentially compromises this aim. Equally importantly, it allows one to obtain a population of T cells that all express the same antigen receptor and that recognise in a highly specific manner a single peptide/MHC complex. T cell clones have therefore been, and continue to be, an invaluable resource for studies on the structure and function of the T cell receptor and the nature and derivation of T cell antigens and epitopes. They have also been frequently used for studying the functional capabilities of T cells, the assumption being that the behaviour of the clonal population exactly reflects that of the original progenitor cell. However, considerable caution needs to be exercised here, as it is clear that the functions displayed by cultured T cells are highly dependent on the culture conditions used, and can easily be affected by culture 'artefacts' such as senescence and mycoplasma infection. It should also be remembered that clones are a highly selected set of cells that in terms of both specificity and function may or may not be representative of the *in vivo* population. As a final caution, it must not be assumed that all members of a clonal population are identical. Apart from the obvious variation associated with the cell cycle or activation status of the cells, heterogeneity in the expression of certain surface markers is not uncommon, and functional studies have revealed a surprising semi-stochastic variability in the responsiveness of individual cells to stimulation.

T cell cloning by limiting dilution

Cloning can be performed either directly with *ex vivo* populations or following enrichment of antigen-specific cells using the methods described above for the derivation of T cell lines. The former method has the advantage that the resulting clones will be as representative as possible of the *in vivo* population. However, it must be borne in mind that because the frequency of antigen-specific T cells in the starting population cannot be determined (except in TCR transgenic mice) the cloning efficiency and therefore the extent to which the derived clones are 'typical' is unknown. The cloning method by limiting dilution provides the further problem that there is always some doubt as to the single cell origin of the clones. Most investigators use Poisson statistics to determine the probability that a 'clone' arose from a single progenitor cell. Typically probabilities of >95% are considered acceptable, a condition that arises when a 96-well cloning plate contains less than 10 positive wells. Since, as already mentioned, neither the frequency of antigen-reactive cells in the starting population nor their cloning efficiency is known, it is necessary to set up cloning plates covering a wide range of input cell numbers and, depending on the number of clones one needs, perhaps several plates at each dose. This is a laborious procedure, but at least the titration of effector cell numbers allows one to check that the limiting dilution cloning procedure conforms to a single-hit process. Although there is always doubt about the monoclonality of clones obtained by limiting dilution, it is likely that even in those cases where the cells in a positive well originate from two progenitor cells, one will be more 'vigorous' than the other and self-cloning will occur. However, in cases where the issue of monoclonality is paramount clones must be recloned, preferably by micromanipulation.

Cloning by micromanipulation

The cloning of enriched populations or lines of T cells can be done by limiting dilution, but if the enrichment is sufficient, cloning by micromanipulation should be considered the first choice. Where the equipment is available, cloning on a cell sorter would be the simplest and most effective method. This latter procedure allows the further possibility of simultaneously selecting the cells to be cloned on the basis of their expression of a variety of surface markers or size parameters.

Expansion of human T cell clones

It is generally assumed that addition of the specific antigen is a prerequisite to generate and expand antigen-specific T cells *in vitro*. As already noted, in the human a serious disadvantage of this method is the requirement for autologous or MHC-matched APC. An alternative approach exploits the fact that once antigen-specific T cells have been enriched in the population by serial antigen-specific stimulation, as described above, clones can be derived and maintained by using PHA as the stimulant and feeder cells, either fresh PBMC or EBV-transformed LCL, derived from an unrelated donor. This method has the further advantage that clones with different antigen specificities can all be stimulated simultaneously with the same feeder cell mix without having to pay any attention to the MHC type of the APC. The combined use of PBMC and EBV-transformed cells as feeders often gives the best result, perhaps because the PHA stimulates the production of growth factors from the PBMC. With this method we have generated $CD4^+$ T cell clones specific for antigens such as tetanus toxoid (Yssel *et al.*, 1986), mycobacterial antigens (Haanen *et al.*, 1991), and $CD8^+$ CTL specific for influenza virus matrix protein (De Waal Malefyt *et al.*, 1993). The polyclonal cloning method we use offers distinct advantages for growth of tumor-specific CTL clones of which the peptide-specificity is not known. We have also generated CD8$^+$ CTL clones from melanoma patients following two stimulations of T cells of patients with autologous melanoma cells. Antigen-independent expansion of the clones was achieved by using as stimulant a mixture of irradiated PBMC, irradiated EBV-LCL and PHA (see the methods section for details). An additional advantage of this approach is that some antigen-specific clones do not grow well in response to their cognate antigen. Ottenhof and co-workers have shown that *M. leprae*-specific T cell clones that did not expand in response to specific antigen, and in fact suppressed the proliferation of other antigen-specific T cell clones (Ottenhof *et al.*, 1986), could be easily expanded by periodical restimulation with feeder cells and PHA.

Growth characteristics of T cell clones

Following stimulation, T cell clones are typically capable of undergoing a 100–1000-fold expansion in the presence of exogenous IL-2. However, whereas mouse T cell clones can often be maintained in culture indefinitely (i.e. appear to be immortal), human T cell clones generally have a maximum life span that approximates the Hayflick limit of 20–25 divisions. Towards the end of their *in vitro* life, which may be considerably shorter that this, a slowing down of their rate of proliferation and degree of expansion following stimulation is noted, often accompanied by loss of function (cytotoxicity, capacity to secrete certain cytokines) and of some surface markers. It has been speculated that the reason for the difference in longevity between mouse and human T cells *in vitro* is the frequent presence of retroviruses in mouse cells and tissues, and that during the early stages of *in vitro* T cell growth one or more of these viruses causes a pseudo-transformation event. Despite this there is a close similarity in the culturability of the different T cell subsets from these two species, CD4 cells being the easiest to grow, CD8 cells being more difficult, and γδ T cells sometimes being very difficult, especially in the mouse. It is unclear as to whether these differences are due to innate differences in the cell biology of different T cell subsets, to factors relating to their life history *in vivo*, or to inadequacies of the culture systems.

In the human, CD8 T cells tend to show lower cloning efficiencies and expansion rates than CD4 T cells. The difference is more noticeable when dealing with antigen-specific T cell clones than non-antigen specific or alloreactive clones, suggesting that antigen-specific memory CD8 T cells may have more limited proliferative capacity *in vitro* than virgin cells. In the mouse, CD8 T cell clones show a tendency to develop into very large granular cells that expand poorly if at all. They also tend to die off much more rapidly than CD4 T cells during the resting phase of the growth cycle. The former problem is minimised by restricting the rate and extent of expansion following stimulation, i.e. using no more than 100 IU/ml IL-2 and terminating the expansion phase within 7 days of stimulation by reducing IL-2 levels to 10–20 IU/ml or (better) by transferring the cells to medium containing 2% rat TCGF and allowing the cells to rest in this medium for at least 7 days.

Mouse $\gamma\delta$ T cell clones are very difficult to grow, some rapidly degenerating into large granular non-proliferative cells, others failing to expand well or respond to restimulation following cloning. They are also susceptible to rapid death in the resting phase and need to be nursed by frequent refeeding with 2% TCGF during this phase. Human $\gamma\delta$ clones vary considerably in their growth rate and stability. $CD4^+$ $\gamma\delta$ clones grow exceptionally well, comparably to $CD4^+$ $\alpha\beta$ T cells (Spits *et al.*, 1991). Some $CD4^-CD8^-$ $\gamma\delta$ clones also grow very well, while others cannot be expanded beyond 10^7 cells. Clones expressing TCR $V\delta2$ gene expand, in general, more easily in our experience than TCR $V\delta1$ expressing T cells. The reason for this is unclear.

A method for *in vitro* expansion of human T cells and generation of cloned T cell lines

Medium to culture T cell clones

For generation and expansion of T cell clones we use Yssel's medium (Yssel *et al.*, 1984). In our hands this medium is superior to RPMI 1640 medium supplemented with either human or fetal calf serum. Yssel's medium is prepared by dissolving the following in the given order in distilled water (suppliers of the ingredients and catalogue numbers are indicated):

1 Iscove's modified Dulbecco's medium (IMDM) (GIBCO, Glasgow, UK, 074-2200).
2 Sodium bicarbonate ($NaHCO_3$), 3.024 g per litre (Merck, Darmstadt, Germany, 6329).
3 Bovine serum albumin (BSA) (Sigma, St Louis, MO, A9647) to a final dilution of 0.25% (w/v).

Table 4.1. *Materials for generation of human T cell cultures*

Reagents for isolation of lymphocytes		
Ficoll Hypaque (lymphoprep)	Nycomed AS	Oslo, Norway
mini MACS (varioMACS)	Miltenyi Biotec	Bergisch-Gladbach, Germany
Fluorescence activated cell sorter	Becton Dickinson	San José, CA, USA
Monoclonal antibodies specific for CD3 (Leu-4), TCR$\alpha\beta$, TCR$\gamma\delta$, CD4 (Leu-3) and CD8 (Leu-2)	Becton Dickinson	San José, CA, USA
Culture media, mitogens and growth factors		
Yssel's medium	(see main text)	
RPMI 1640 (12-702B)	BioWhittaker	Walkersville, MD, USA
PHA (H 16/17)	Wellcome	Beckenham, UK
Recombinant IL-2 (Proleukin)	Chiron	Amsterdam, Netherlands
Tissue culture plates		
96 round-bottomed well plate (3799)	CorningCostar	Acton, MA, USA
24-well plate (3524)		
Various tissue culture flasks (for culturing EBV-LCL)		

4 2-amino ethanol (Merck 102931), final concentration 1.8 μg per litre.

5 Transferrin (Boehringer Mannheim, Germany, 652-202) liquid, final concentration 40 μg per ml.

6 Insulin (Sigma, I 5500) to a final concentration of 5 μg per ml. (dissolve insulin 10 mg/ml in 0.01 M HCL and add this to the medium).

7 Linoleic acid (Sigma, L 1376) to a final concentration of 2 μg per ml.

8 Oleic acid (Sigma, O 3879) to a final concentration of 2 μg per ml (stock linoleic and oleic acid should be stored at $-20\,°C$ under nitrogen to prevent oxidation of unsaturated bonds. Therefore one should make glass ampoules containing 5 mg of linoleic and 5 mg of oleic acid. For preparation of the medium, dissolve the fatty acids in ethanol and add this mixture to the medium).

9 Antibiotics (penicillin, 100 U/ml and streptomycin 100 μg/ml).

Filter the medium (0.45 μ filter is sufficient), aliquot, and store at $-20\,°C$. The medium can be stored for up to 4 months at $-20\,°C$. After thawing one should not keep the medium through more than 7 days at $4\,°C$.

Note

Instead of powdered medium, one can also use liquid IMDM (GIBCO, 041–90560 M). Although many cell types can be cultured in Yssel's medium, addition of 1–2% pooled human serum (collected from the blood of 6–19 healthy donors and decomplemented by putting the tube or flask with serum for 30 min in a waterbath at 56 °C) is recommended.

In order to save time in preparation of the medium, a $5 \times$ concentrated stock solution in IMDM of items 3–10 can be made and stored at -20 °C, which can be diluted to obtain $1 \times$ medium.

Preparation of feeder cell mixture

PBMC are isolated from freshly obtained peripheral blood by centrifugation over Ficoll–Hypaque (Lymphoprep, Nycomed Pharma AS, Oslo, Norway). The PBMC are irradiated with 40 Gy, spun for 10 min at 1400 rpm in a standard table-top centrifuge and resuspended in Yssel's medium (see below) at a concentration of $10–20 \times 10^6$ cells per ml. The EBV-LCL are irradiated with 50 Gy, spun down and resuspended in Yssel's medium at a concentration of $1–2 \times 10^6$ cells per ml. With our feeder cell mixture, consisting of irradiated PBMC, EBV-LCL and PHA (Wellcome, see below), it is possible to expand freshly isolated T cells to large numbers (see section about expansion of fresh T cells and clones). Usually the cultures are carried out in 24-well flat-bottomed plates (Falcon).

Cloning of T cells

We have employed two methods of T cell cloning. One method uses limiting dilution of the T cells and the second uses single-cell cloning with the FACStar plus (Becton Dickinson, San José, CA). For cloning by limiting dilution we use the following procedure: A feeder cell mixture is made up consisting of 5×10^5 irradiated PBMC, 5×10^4 irradiated EBV-LCL and 50 ng PHA (make up stock PHA at 10 µg/ml; purified PHA is from Wellcome Diagnostics, Beckenham, Kent, UK, HA 16/17) per ml. Next, serial dilutions of 10^5, 10^4, $10^{3,}$ 10^2, 10 and 1 T cells per ml are made with the feeder cell mixture as diluent and 100 µl aliquots are pipetted per well of a 96-well, round-bottomed microtitre tray (CorningCostar, Acton, MA, USA). The number of wells that should be filled with each dilution depends upon the expected cloning efficiencies.

Alternatively, T cells can be cloned with a cloning device coupled to the FACStar plus (Becton Dickinson, San José, CA). The cells that need to be cloned are labelled with FITC and/or Phycoerythrin (PE)-labelled antibodies, washed at least twice, and subjected to FACS cloning. This method has the advantage that very small subsets can be cloned without prior enrichment. For example, if one wishes to have clones of cells that constitute less than 0.1% of total PBMC, those cells can be directly cloned provided one has mAbs detecting that subset. Since it can take a long time to clone very rare subsets with the FACStar plus, it is advisable to enrich the cells of interest before cloning. For example, in our experiments to clone $CD4^+TCR$ $\gamma\delta^+$ cells (<0.1% of total PBMC), we first enriched the total TCR $\gamma\delta^+$ population with magnetic bead sorting before labelling with anti-CD4 mAb. For this the miniMACS system of Miltenyi (varioMACS, Miltenyi Biotec, Bergisch Gladbach, Germany) can be used, following the instructions of the manufacturer. The number of cells added per well of a microtitre tray by the FACS can be adjusted. The numbers of cells added per well and the number of filled wells depends on the expected cloning efficiencies. For example, cloning efficiencies of 40–80% can be expected when one clones $CD4^+$ T cells. To obtain 50 clones it may be sufficient to fill one or two trays with one single cell per well. The cells are collected in round-bottomed wells (96-well plates can be purchased from various companies; we use Costar) filled beforehand with 100 µl feeder cell mixture.

One week after cloning, 100 µl of medium containing 20 U per ml recombinant IL-2 (Chiron, Amsterdam, Netherlands) is added. Growing cultures of T cells should become visible over the next 2–7 days. Growing cultures are usually transferred between days 14 to 21 to wells of a 24-well plate and are then restimulated with feeders for expansion (see next paragraph) in a total volume of 1 ml per well.

Expansion of freshly isolated T cells or clones with feeder cells

Make up 2 × concentrated feeder cell mixture (A) as follows: PBMC, 2×10^6 cells/ml; EBV transformed B cells, 2×10^5 cells/ml; PHA, 0.2 µg per ml (make up stock PHA at 10 µg/ml, purified PHA from Wellcome HA 16/17).

Wash the T cells twice in Yssel's medium and resuspend at a concentration of 4×10^5 cells/ml (B). Add 0.5 ml of feeder cell mixture (A) and 0.5 ml of fresh T cells or T cell clone suspension (B) to each well. Final concentrations are 2×10^5 cells per ml for the T cells and for the feeder cell mixture:

PBMC, 10^6 cells/ml; EBV-LCL, 10^5 cells/ml and PHA, 0.1 µg/ml. The concentrations of T cells and feeder cells are chosen in such a way that during the first 3–4 days after stimulation, the cultures should grow without intervention. At day 4 after stimulation, the cultures are subcultured into Yssel's medium containing 20 U ($\pm$ 2 ng) rIL-2 per ml. Vigorous growth should be observed during the first 7 days after restimulation with feeder cells, but it should be realised that the growth rates of cloned lines may differ from one line to another.

Feeder cells usually disappear between days 4 and 7 after restimulation. Very fast growing cell lines may 'clear' the feeder cells more rapidly. Slow-growing T cells (doubling times 3 days or more) do not clear the feeder cells and it is advisable to spin the cells over a Ficoll–Hypaque gradient to remove dead cells before using the T cells in assays. If one has 10^5–10^6 cells these may be layered in a volume of 2 ml over 3 ml of Ficoll–Hypaque in 15 ml conical tubes (higher numbers of cells should be layered over 10 ml of Ficoll–Hypaque in 50 ml conical tubes), followed by centrifugation at 2000 rpm. The interphase is removed and washed three times. Optimal growth rates are achieved when the cell concentration is maintained at $\pm 10^6$ cells/ml. It should however be noted that the T cells can sustain very high concentrations (up to 9×10^6 cells/ml) provided that the medium is not exhausted. T cells have to be restimulated periodically for expansion. The optimal time for restimulation varies depending on the kind of T cells. Most TCR $\gamma\delta^+$ and $CD8^+$ T cells and some $CD4^+$ T cell cultures should be restimulated at 7 day-intervals. Most $CD4^+$ T cell clones, some well-growing TCR $\gamma\delta$ and $CD8^+$ T cell clones can be restimulated at 10–14 day intervals. While for some clones the restimulation interval is critical (these cells die rapidly if they are not restimulated) other clones can survive prolonged periods of time. In particular some $CD4^+$ T cell clones can survive culture periods of over two months without restimulation, provided that the medium is refreshed regularly.

It is very important to check the cultures regularly for mycoplasma as this insidious infection severely affects growth and functional properties of T cells. This holds not only for the cultures of T cells but of course also for the EBV-transformed B cells used as feeder cells. As mycoplasma infection is not visible, it may spread unnoticed when cultures are not monitored regularly. This is especially disastrous if one has made frozen stocks of cultures that are mycoplasma positive. In our laboratory, testing is done once in 14 days using a mycoplasma detection kit (Gen-Probe Inc., San Diego, CA). It needs to be borne in mind that mycoplasma is not sensitive to most antibiotics, therefore one may omit antibiotics altogether. Appearance of bacterial or fungal infections is the clearest sign that the procedures of culture are not properly exercised and are often predictive of mycoplasma infections.

Freezing of cultured T cells

Cultured human T cells are quite robust and a simple method to freeze these cells can be used. We freeze the cells as follows: spin the cells at 250g for 5 min at room temperature and resuspend in RPMI-1640 medium supplemented with 10% fetal bovine serum. Put the cells on ice and prepare freezing medium (RPMI–10% fetal bovine serum plus 20% DMSO). Let the freezing medium cool on ice and mix the cell suspension with an equal volume of 20% DMSO. Distribute the cell suspension into 1 ml freezing tubes and place them in a styrofoam box. Place the box in a −70°C freezer for two days. The tubes can be stored in the vapour phase of liquid nitrogen. For thawing, quickly warm the tubes in a 37°C water bath, mix 1 part of cell suspension with 10 parts RPMI medium and spin for 5 min at 250g at room temperature followed by two washes. The recovery is mostly around 50% of the input. Procedures such as computer programmed freezing or careful dropwise dilution of the cell suspension after thawing do not improve the cell recovery.

A method for the generation of mouse T cell clones

The method described here is for the generation and maintenance of allospecific mouse T cell clones. It can be readily adapted to the generation of other types of clones. For instance, to generate non-antigen-specific T cell clones the principal differences would be (a) the use of syngeneic rather than allogeneic feeder cells; (b) the addition of ConA at 3 µg/ml to stimulation cultures. To generate antigen-specific CD4 T cell clones the differences would be: (a) responder cells would be obtained from mice that had been immunised with the relevant antigen; (b) the use of syngeneic feeder cells; and (c) the addition of soluble antigen at an optimal concentration to stimulation cultures. Details of the general equipment and materials needed and the method of preparing spleen cells are given in the description of the culture of mouse NK cells in Chapter 7. Mouse T cell clones can be generated using almost any complete medium. It is most common to use RPMI 1640, but in our hands better results, especially with difficult clones, are obtained with high-glucose DMEM or Click's medium. Commercially prepared medium is generally adequate, but for the most critical applications (e.g. TCR γδ T cells) we prefer to use medium made up from powdered stocks with Nanopure water (see Chapter 7). It is essential to supplement media with serum, usually 10%.

Preparation of TCGF (ConA-induced rat spleen cell supernatant)

Spleen cells are prepared from 5–10 spleens obtained from young adult rats in an identical manner and using the same reagents as described for the preparation of mouse spleens in Chapter 7. The rat strain used is probably unimportant but we have always used Wistars. It is best to prepare the cells from each spleen separately, pooling the cells together in a 50 ml centrifuge tube after the first wash, and washing two more times. All washes should be performed at 4°C. After the last wash cells are resuspended in RPMI 1640 supplemented with 5×10^{-5} M 2-mercaptoethanol and 10% fetal bovine serum and adjusted to 5×10^6/ml.

ConA (type IV, Sigma 2010) is added to give 10 µg/ml, and 50 ml aliquots are added to 75 cm^2 flasks. The flasks are incubated horizontally at 37°C, and after 40 h the supernatant is harvested and spun at 1000g for 20 min to remove cells and debris. Clear supernatant from the spun tubes should be carefully collected by pipette and filtered first through an 0.45 µm filter then an 0.2 µm filter. It can be stored at 4°C indefinitely. Sometimes a cloudiness or precipitate develops with time, but if desired this can be removed by re-centrifugation and/or re-filtering.

Feeder cell mixture

Stimulator/feeder mouse spleen cells are prepared, resuspended in complete culture medium. Just prior to use they are exposed to 20 Gy of X-ray or gamma-ray irradiation and adjusted to 5×10^6/ml in complete medium containing 200 IU/ml recombinant human IL-2.

Cloning of T cells

Spleen cells from the responder strain of mice are prepared in complete culture medium. If one wishes to restrict the developing clones to CD4 or CD8 the spleen cell suspension should be depleted of the relevant population by treatment with Ab and complement as described in Chapter 7. Serial two- or three-fold dilutions of responder spleen cells are made to cover a suitable range of responder input cell numbers per well (for alloreactive clones this would be around 100–1000 cells/well). These are mixed with an equal volume of feeder cell mixture and 100 µl aliquots are added to flat-bottomed 96-well plates. We prefer to use half-area plates as this greatly facilitates the screening for clones at the end of the incubation period. Plates are incubated at 37 °C. Best results are obtained by placing these in a separate humidified (but not air-tight) box at the back of the incubator, and all temptation to remove the plates for examination should be resisted until they have been cultured for 7 days. Because clones tend to develop at very different rates, plates should be screened for colonies on at least three occasions (e.g. days 7, 10 and 14). Promising clones are transferred to 24-well plates with 1 ml medium containing 100 IU/ml IL-2. By day 14 the frequency of clones in each plate will be known and those that had a probability of monoclonality of > 95% are retained. Three to 4 days after transfer to 24-well plates, clones should be refed with 1 ml medium containing 100 IU/ml IL-2 and subcultured if necessary. Seven days after transfer, and again after 10–11 days clones should be refed with medium containing 2% TCGF. By this time clones should have entered the resting state. Some cell death virtually always occurs at this stage, but as long as a reasonable number of viable cells remain there should be no problem in restimulating the clone.

Maintenance of T cell clones

All mouse T cell clones should be maintained on a strict cycle of stimulation, expansion, and rest. The expansion phase in total, from the day of stimulation, should not last more than 7 days, and the rest period, which begins 7 days after stimulation, should not be shorter than 7 days.

To restimulate T cell clones the culture volume in the wells should be reduced by aspiration to about 1 ml and 1 ml of feeder cell mixture added. After 3–4 days the stimulated and now expanding cells should be resuspended, transferred to centrifuge tubes, and underlain with about 1 ml of Ficoll–Hypaque using a Pasteur pipette. The tubes are spun at 400g for 10 min at room temperature using slow acceleration and no brake, and the viable proliferating T cells collected from the interface. These are washed once in 3–5 ml Hanks' medium, resuspended in an appropriate volume of culture medium containing 100 IU/ml IL-2, and 2 ml aliquots added to an appropriate number of wells in a 24-well plate. This number should be judged from experience such that by day 7 after stimulation these expansion wells have a high density of T cells but have not become overgrown.

At day 7 the resting phase should be imposed by removing half of the medium from the wells and adding 1 ml of medium containing 2% TCGF (medium containing 10 IU/ml IL-2 can be used instead, but this gives inferior results for CD8 and γδ clones). Wells should be refed with 1 ml of the same medium every 3–4 days. CD4 T cells can often remain highly viable in the resting state for weeks or months. Cells should not be restimulated until they have been rested for at least 7 days.

At the earliest opportunity T cell clones that are needed for future studies should be frozen down and stored in the vapour phase of liquid nitrogen. The details of the cell freezing method we use are given in Chapter 7. In the case of T cell clones the cells are best harvested for freezing 1–2 days after restimulation, and upon thawing should be placed in a well with 2 ml of medium containing 100 IU/ml IL-2. They should be refed/subcultured with this medium after 2–3 days, and put into the resting state following a total of 7 days of culture after stimulation.

References

Bakker, A. B., Marland, G., de Boer, A., Huijbens, R. J., Danen, E. H., Adema, G. J. & Figdor, C. G. (1995). Generation of antimelanoma cytotoxic T lymphocytes from healthy donors after presentation of melanoma-associated antigen-derived epitopes by dendritic cells in vitro. *Cancer Res.*, **55**, 5330–4.

Cantrell, D. A. & Smith, K. A. (1984). The interleukin-2 T cell system: a new cell growth model. *Science*, **224**, 1312–16.

Celis, E., Tsai, V., Crimi, C., DeMars, R., Wentworth, P. A., Chesnut, R. W., Grey, H. M., Sette, A. & Serra, H. M. (1994). Induction of anti-tumor cytotoxic T lymphocytes in normal humans using primary cultures and synthetic peptide epitopes. *Proc. Natl. Acad. Sci. USA*, **91**, 2105–9.

De Waal Malefyt, R., Verma, S., Bejarano, M., Ranes-Goldberg, M., Hill, M. & Spits, H. (1993). CD2/LFA-3 or LFA-1/ICAM-1 but not CD28/B7 interactions can augment cytotoxicity by virus specific CD8+ CTL. *Eur. J. Immunol.*, **23**, 418–26.

Gotch, F., McMichael, A. & Rothbarth, J. (1988). Recognition of influenza A matrix antigen by HLA-A2-restricted cytotoxic T lymphocytes. *J. Exp. Med.*, **168**, 2045–57.

Haanen, J. B., de Waal Malefyt, R., Res, P. C. M., Kraakman, E. M., Ottenhof, T. H. M., De Vries, R. R. P. & Spits, H. (1991). Selection of a human Th1-like T cell subset by mycobacteria. *J. Exp. Med.*, **175**, 583–92.

Koenig, S., Earl, P., Powell, D., Pantaleo, G., Merli, S., Moss, B. & Fauci, A. S. (1988). Group-specific, major histocompatibility complex class I-restricted cytotoxic responses to human immunodeficiency virus 1 (HIV-1) envelope proteins by cloned peripheral blood T cells from an HIV-1-infected individual. *Proc. Natl. Acad. Sci. USA*, **85**, 8638–42.

Meuer, S. C., Cooper, D. A., Hodgdon, J. C., Hussey, R. E., Fitzgerald, K. A., Schlossman, S. F. & Reinherz, E. L. (1983). Identification of the receptor for antigen and major histocompatibility complex on human inducer T lymphocytes. *Science*, **222**, 1239–42.

Ottenhof, T. H. M., Elferink, D. G., Klatser, P. R. & De Vries, R. R. P. (1986). Cloned suppressor T cells from a lepromatous leprosy patient suppress *Mycobacterium leprae* reactive helper T cell clones from leprosy patients. *Nature*, **322**, 462–4.

Plebanski, M., Allsopp, C. E., Aidoo, M., Reyburn, H. & Hill, A. V. (1995). Induction of peptide-specific primary cytotoxic T lymphocyte responses from human peripheral blood. *Eur. J. Immunol.*, **25**, 1783–7.

Spits, H., Breuning, M., Ivanyi, P. & De Vries, J. E. (1982). In vitro isolated human cytotoxic T-lymphocyte clones detect variations in serologically defined HLA antigens. *Immunogenetics*, **16**, 503–12.

Spits, H., Paliard, X., Vanderkerckhove, Y., Van Vlasselaer, P. & De Vries, J. E. (1991). Functional and phenotypic differences between CD4$^+$ and CD4$^-$ TCR $\gamma\delta$ clones from peripheral blood. *J. Immunol.*, **147**, 1180–8.

Van der Bruggen, P., Bastin, J., Gajewski, T., Coulie, P. G., Boël, P., De Smet, C., Traversari, C., Townsend, A. & Boon, T. (1994). A peptide encoded by human gene MAGE-3 and presented by HLA-A2 induces cytolytic T lymphocytes that recognize tumor cells expressing MAGE-3. *Eur. J. Immunol.*, **24**, 3038–43.

Van Pel, A., Van der Bruggen, P., Coulie, P. G., Brichard, V. G., Lethe, B., Van den Eynde, B., Uyttenhove, C., Renauld, J. C. & Boon, T. (1995). Genes coding for tumor antigens recognized by cytolytic T lymphocytes. *Immunol Rev.*, **145**, 229–50.

Yssel, H., De Vries, J. E., Koken, M., van Blitterswijk, W. & Spits, H. (1984). Serum-free medium for the generation and the propagation of functional human cytotoxic and helper T cell clones. *J. Immunol. Methods*, **72**, 219–27.

Yssel, H., Blanchard, D., Boylston, A., De Vries, J. E. & Spits, H. (1986). T cell clones which share T cell receptor epitopes differ in phenotype, function and specificity. *Eur. J. Immunol.*, **16**, 1187–94.

5

B lymphocytes

Anne E. Corcoran and Ashok R. Venkitaraman

Introduction and potential research applications

It is essential to view techniques for B lymphocyte culture in the light of remarkable recent progress in elucidating the molecular basis of the development of this lineage. Two processes are fundamental to development: the **rearrangement** of germline DNA encoding the immunoglobulin heavy (IgH) and light (IgL) chains, and the waves of cellular **proliferation**, which must be coordinated with Ig gene rearrangement. In this section, we briefly review current understanding of the mechanism and control of these processes.

The endpoint of the early steps in B lymphopoiesis is a cell that can recognise and respond to foreign antigens through an antigen receptor, the B cell receptor or BCR (Fig. 5.1). Antigen recognition by the BCR is mediated by membrane-inserted Ig molecules, of the IgM or IgD class in naive B cells, or of other Ig classes in antigen-primed cells. In addition, the BCR also contains at least two co-receptor molecules, Igα and Igβ, which are essential for transmembrane signalling in response to antigen. Recognition of diverse antigen specificities by the humoral immune response dictates that a vast array of Ig molecules be encoded in the genome.

The genes encoding Ig molecules are unique in their segmental organisation in the genome. Thus the DNA encoding the variable, N-terminal domain of IgH and IgL chain polypeptides is arranged in 2–3 discrete segments (termed V, D and J) in the germline (Fig. 5.1). During B lymphocyte development in the bone marrow, these segments are brought together by a series of DNA rearrangements to give rise to functional Ig genes. The rearrangement process contributes, in several ways, to the enormous diversity of the variable domains of Ig molecules. Numerous V

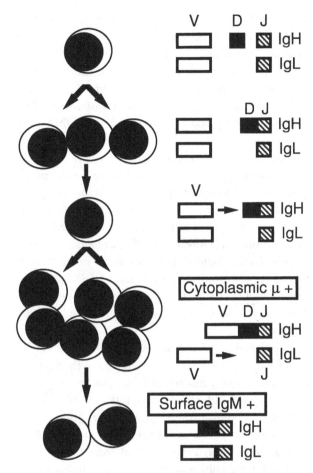

Fig. 5.1. Cell proliferation accompanies gene rearrangement during B lymphopoiesis. Proliferative stages are indicated on the left, and steps in Ig gene rearrangement on the right.

segments are found in the *IgH* and *IgL* loci, generating combinatorial diversity by their random usage in functional Ig genes. Re-joining of segments during rearrangement is imprecise, creating junctional diversity. Finally, functional Ig genes in mature B lymphocytes undergo somatic hypermutation during a secondary immune response, resulting in the creation and selection of antibodies with ever-higher affinity for a given antigen.

Rearrangement involves the creation and repair of double-strand DNA breaks at conserved signal sequences flanking each segment (Schatz,

Oettinger & Schlissel, 1992). Only two components of the recombinase machinery that carries out these steps are specific to lymphoid cells – the recombinase activating genes *RAG1* and *RAG-2*. *RAG-1* and *RAG-2* are together sufficient to carry out cleavage at the appropriate sites in DNA, whilst ubiquitous molecules mediate the repair steps. In addition, the lymphoid-specific enzyme terminal deoxynucleotide transferase (TdT), contributes to junctional diversity by adding bases to cleaved DNA ends prior to rejoining.

The rearrangement of Ig genes is an ordered process (Schatz *et al.*, 1992). The *IgH* loci are the first to rearrange during B lymphocyte development, beginning with D to J joining and followed by V to (D)J. V to J joining in the *IgL* loci generally occurs after functional *IgH* rearrangement. The mechanisms that initiate Ig gene rearrangement in progenitors committed to the B lymphocyte lineage remain unclear. However, feedback regulatory signals following the initiation of *IgH* rearrangement are delivered by its own partial products. Thus developing B cells monitor the progress of *IgH* rearrangement by sensing the transmembrane signals delivered by the membrane-inserted products of (D)J as well as V to (D)J joining in association with the co-receptor proteins $Ig\alpha$ and $Ig\beta$.

Signalling through the so-called pre-B cell receptor (pre-BCR) is particularly important to the normal progression of Ig rearrangement (Melchers *et al.*, 1994). The pre-BCR consists of a functional IgH chain in association with surrogate light chains (VpreB and λ5) and $Ig\alpha/Ig\beta$. Assembly of the pre-BCR requires the successful completion of V(D)J rearrangement in at least one *IgH* allele, and the synthesis of a membrane-inserted heavy chain of the IgM class (μm). Signalling through the pre-BCR stops further rearrangement on the second *IgH* allele, a phenomenon termed allelic exclusion, ensuring that each mature B lymphocyte contains an antigen receptor with a single specificity. Pre-BCR assembly also signals to increase the frequency of *IgL* rearrangement, helping to maintain the order in which the loci undergo rearrangement.

Because each joining event is imprecise, and therefore often ineffectual, rearrangement leads to the successful production of a functional Ig gene in only a minority of all attempts. Rearrangement is therefore accompanied by waves of cell proliferation, ensuring an adequate throughput of mature B cells to the periphery. An essential signal for proliferation is delivered by the cytokine interleukin-7 (IL-7) secreted by bone marrow stromal cells, which binds to its heterodimeric receptor (IL-7R) expressed on the surface of B cell progenitors. The IL-7R includes a ligand-specific α chain, as well as a γ

chain shared with other cytokine receptors. The interaction of IL-7 with the IL-7R is essential for normal B lymphocyte development. Thus, targeted disruption of the gene encoding IL-7, or of either chain of the IL-7R, results in a severe impediment to development. In addition to its role in triggering cell growth or survival, the IL-7R also regulates differentiation (Corcoran *et al.*, 1996), although the mechanism by which it does so has not been clarified at the molecular level.

It must be emphasised that the IL-7/IL-7R interaction does not provide the only signal for proliferation during B lymphopoiesis. In addition, contact with stromal cells is essential for cell division at very early stages, and may work through the interaction of the transmembrane form of the growth factor SCF on stromal cells with its cognate ligand, c-kit, on lymphocyte progenitors. Insulin-like growth factor 1, the chemokine SDF and the cytokine IL-3 have also been observed to cause the proliferation of progenitor B cells although their physiological role is yet to be precisely determined. A recently identified cytokine, thymic stromal derived lymphopoietin (TSLP) has *in vitro* effects that mimic those of IL-7: it is even believed to bind to a multimeric receptor that shares the IL-7R α chain. However, the severity of the developmental block in the absence of IL-7 suggests that the requirement for TSLP is less stringent.

Characterisation and markers of the B lymphocyte lineage

Murine B lymphocytes

It has by now become apparent that B lymphocyte development in the murine bone marrow proceeds in discrete stages that can be distinguished on the basis of markers expressed at the cell surface. Identification and isolation of cells at these stages has been of great importance to techniques in B lymphocyte culture. Alternative systems of nomenclature have been proposed by the two groups that have made significant contributions in this area. Both systems will be considered in this discussion, together with the unifying principles that link them.

Three broad stages can be identified in B lymphocyte development based on the state of rearrangement of the *IgH* and *IgL* loci (Figs. 5.1 and 5.2). Progenitor, or **pro-B cells** have initiated but have not yet completed *IgH* rearrangement. Successful *IgH* rearrangement and the assembly of the pre-BCR triggers developmental progression to the precursor, or **pre-B cell** stage. This is followed by *IgL* rearrangement, leading to the generation of **B lymphocytes** that express an IgH/IgL antigen receptor at their surface.

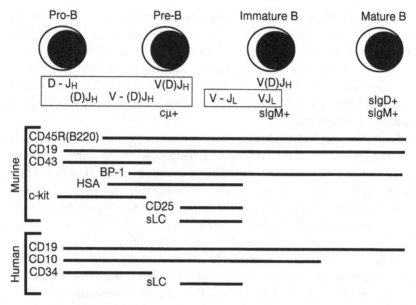

Fig. 5.2. Surface markers expressed during stages of B lymphopoiesis in murine and human bone marrow. Note that Ig gene rearrangements occur as a continuum across developmental stages. cμ, cytoplasmic μ heavy chain. sLC, surrogate light chain.

The surface marker CD45R (B220), an isoform of the transmembrane phosphatase CD45 restricted to B lymphocytes, has long been used as a lineage marker (Fig. 5.2). Recent evidence suggests, however, that B220 is also expressed in cells belonging to the T cell and natural killer cell lineages (Rolink et al., 1996). Therefore, the marker CD19, an accessory molecule of the BCR, is increasingly used as an alternative to B220 (Fig. 5.2). CD19 expression is restricted to B lymphocytes, and is maintained from very early stages in development through to maturity.

Pro-B cells co-express the B220 and CD19 lineage markers with CD43 (Hardy et al., 1991; Li, Hayakawa & Hardy, 1993). They can be divided into distinct subsets (Fig. 5.2) based on differential expression of two additional molecules, the markers Heat Stable Antigen (HSA) and BP-1. The earliest pro-B cells (termed Fraction A by Hardy) express neither HSA nor BP-1. HSA expression is initiated in pro-B cells belonging to Fraction B, and increases in Fraction C, at which stage BP-1 is also detected. A continuum of IgH rearrangements is found during these stages (Li et al., 1993; Ehlich et al., 1994). IgH rearrangements are virtually undetectable in Fraction A cells; whilst Fractions B and C are marked by a gradual increase in D–J joins progressing smoothly to the onset of V to (D)J rearrangements (Fig. 5.2).

Melchers, Rolink and co-workers (ten Boekel, Melchers & Rolink, 1995) distinguish pro-B cells by differential expression (Fig. 5.2) of c-kit, the SCF receptor, and the marker CD25, encoding the α subunit of the IL2 receptor. Pro-B cells belonging to Hardy's Fractions A–C can be identified as $CD19^+ckit^+CD25^-$ in this system of nomenclature.

The expression of CD43 and c-kit is extinguished (Fig. 5.2) as pro-B cells successfully complete V(D)J rearrangement on at least one *IgH* allele and progress to the pre-B cell stage. This is accompanied by the onset of CD25 expression. Thus pre-B cells are defined in the Melchers & Rolink system to be $CD19^+c\text{-}kit^-CD25^+$, whilst Hardy's Fraction D is characterised as $CD19^+CD43^-BP\text{-}1^+HSAhi$. However defined, these cells contain cytoplasmic μm chains reflecting the successful completion of *IgH* rearrangement.

IL7-dependent proliferation occurs at the pre-B cell stage, with its cessation accompanied by the onset of *IgL* rearrangement (Figs. 5.1 and 5.2). Thus, markers of cell cycle status such as cell size, DNA content and the expression of proliferating cell nuclear antigen (PCNA) distinguish large, dividing pre-B cells from small, nondividing ones. It is the latter fraction in which the majority of *IgL* rearrangement is believed to occur. If successful, *IgL* rearrangement leads to the development of B lymphocytes expressing the complete BCR. Initially, these BCRs contain Ig molecules exclusively of the IgM class defining immature B lymphocytes of Hardy's fraction E. Co-expression of IgD with IgM in Hardy's Fraction E (Hardy *et al.*, 1991; Li *et al.*, 1993) marks developmental progression to maturity (Fig. 5.2).

Little is known of the dynamics of mature B lymphocytes that leave the bone marrow and migrate to the periphery. It is clear that cells recruited to the periphery populate lymphoid follicles in secondary lymphoid tissues such as the spleen and lymph nodes, where they come into contact with antigens and interacting cells such as T lymphocytes. Whilst certain antigens may activate B lymphocytes without the helper functions delivered by T cells, most require T cell help. T cell dependent B cell activation leads to terminal differentiation to the antibody secreting plasma cell stage. Moreover, a small fraction of activated B cells undergoes differentiation into long-lived memory cells that are capable of being reactivated upon re-challenge with the same antigen. It is believed that somatic hypermutation of Ig genes and class switching from IgM/IgD to IgG, IgE and IgA take place in memory cells, giving rise to the features characteristic of the secondary immune response.

Human B lymphocytes

There are clear differences between the development of B lymphocytes in humans and mice, as well as obvious similarities. The enzymatic machinery mediating gene rearrangement is virtually identical in both species, although the Ig loci which are their substrates exhibit some differences. For instance, whilst κ light chains account for 95% of all murine light chains, the $\kappa:\lambda$ ratio of human antibodies is close to 1:1 and the human λ locus is correspondingly more diverse.

The surface markers used to distinguish successive stages in human B lymphocyte development (Fig. 5.2) are somewhat different from those used in murine cells (Ghia et al., 1996). Coexpression of CD19 with CD10 identifies bone marrow B lineage progenitors. $CD19^+CD10^+$ cells can be further separated into $CD34^+$ cells undergoing $(D)J_H$ and V to $(D)J_H$ rearrangements, then $CD34^-preBCR^+$ (cytoplasmic μ^+) cells, which have successfully completed IgH rearrangements, followed by the appearance of $sIgM^+$ immature B cells expressing a surface BCR. Extinction of CD10 expression accompanies maturation to $sIgM^+sIgD^+$ mature B lymphocytes. The temporal order of rearrangements in the human Ig loci is similar to that during murine B lymphopoiesis (Ghia et al., 1996).

Perhaps the most fundamental difference between human and murine B lymphopoiesis pertains to the signals that trigger cellular growth and differentiation. Insights into these differences have come from the study of human immunodeficiency diseases in which B lymphopoiesis is perturbed. In contrast to mice, deficiency of the shared cytokine receptor γ chain in humans (X-linked severe combined immunodeficiency) does not greatly affect B cell development, implying that the IL-7/IL-7R interaction is not essential. Conversely, mutations of the Btk tyrosine kinase activated by signalling through the pre-BCR result in a severe block at the pre-B cell stage in X-linked or Bruton's agammaglobulinaemia in humans, but not in mice.

Overview of methods for the culture of early stages in B cell development

a. Adult bone marrow

The first long-term murine bone marrow cultures (LTBMC), with which their names have become synonymous, were established and characterised by Whitlock & Witte (1987). The protocols given below are based primarily on their work, with later modifications introduced by others. Whitlock

& Witte established that bone marrow lymphocytes cultured *in vitro* in the presence of a stromal cell layer could be maintained and expanded for weeks or months. Progression through several B cell developmental stages could be observed. A number of factors were found to be critical for successful cultures including: cell density (cultures could not be initiated at a cell density less than 10^6 per ml, while too great a density caused overgrowth by macrophages); stromal feeder cell layer (pre-established at a cell density of 3×10^5/ml); fetal calf serum (large batch-to-batch variation was found in the ability of FCS to support B cell expansion partly due to inhibitory presence of glucocorticoids, with serum concentrations greater than 5% inhibiting B lymphopoiesis and encouraging non-stromal adherent cell outgrowth); mouse strain and age (cultures were particularly successful in the BALB/c background, using mice aged 3–4 weeks as a source of bone marrow).

Nishikawa and co-workers established that the stages of B cell development in murine bone marrow can be defined by their requirement for stromal cells and IL-7. The very earliest precursors adhere to and are dependent on the stromal layer for expansion. In the next developmental stage when pro-B cells undergo DJ rearrangements, IL-7, in addition to the stromal layer, is an absolute requirement for growth. Upon completion of functional IgH rearrangements and progression to the pre-B cell stage, cells no longer require the stromal layer, but still require IL-7 and the co-stimulatory growth signal supplied by the pre-BCR (Hayashi *et al.*, 1990).

The availability of recombinant IL-7 has greatly aided the establishment of successful B cell cultures. Nishikawa and colleagues have designed a defined medium including IL-7 that will support the growth of progenitors without the need for stromal cells (Yasunaga *et al.*, 1995). However, these cultures do not support differentiation into more mature B cells. They, and others, have also characterised a number of cloned stromal cell lines that can substitute for primary stromal layers in supporting B lymphopoiesis.

Rolink, Melchers and colleagues have exploited the selective expression of c-kit by murine pro-B cells to establish long-term cultures. Unfractionated bone marrow is sorted to enrich for the B220[+], c-kit[+] fraction, which is then cultured with the stromal cell line PA6 (described below) and IL-7. Under these conditions the cultures grow for months and can be cloned at high plating efficiencies (Rolink *et al.*, 1993). In this culture system, pro-B cells do not spontaneously complete V(D)J rearrangement, but can be induced to do so by the removal of IL-7. In this respect, the use of purified progenitor B cells differs from cultures using unfractionated bone marrow, in which IL-7 supports the outgrowth of cytoplasmic μ expressing cells (Hayashi et al., 1990). These differences may be related to a requirement for primary stromal cells, or an as yet unknown accessory cell, or their products.

Long-term cultures of human bone marrow B lymphocytes have proved difficult to establish. Cooper and colleagues have cultured fetal or bone marrow derived mononuclear cells in culture medium containing IL-7 for up to 3 weeks, during which proliferation and differentiation can be observed (Nemerow, McNaughton & Cooper, 1985). Rolink and colleagues report some success with short-term (up to 3 weeks) culture and in vitro differentiation of the sorted CD19$^+$, CD10$^+$, κ- and λ – precursor population (Ghia et al., 1995). These require a stromal cell line, either human or murine. Some cells survive for analysis but very little cell expansion occurs.

b. Early B cell differentiation in the murine embryo

The fetal liver is the site of embryonal B cell differentiation. B cell progenitors begin to appear at Day 13 of gestation, developing into pre-B cells, which undergo heavy and light chain rearrangement and express surface Ig from Day 16, at which stage they become stromal cell independent. Coculture of Day 13 fetal liver cells with a stromal cell layer allows them to differentiate from B220$^-$ cells to B220$^+$, IL-7 responsive cells, and to progress to later stages in development.

c. B-I B cells

In addition to 'conventional' or B-2 B cells, another type of B cell – the B-1 B cell – has a distinct tissue distribution in the mouse, and is thought to belong to a distinct lineage. B-1 B cells develop in the fetal liver. In the adult they are found primarily in the peritoneal cavity, with a small number in the spleen but virtually none in the bone marrow and lymph nodes (in contrast to B-2 B cells). B-1 cells are CD43$^+$, B220^{lo}, IgM^{hi}, IgD^{lo}, CD23$^-$, whereas B-2 B cells are CD43$^-$, B220^{hi}, IgM^{lo}, IgD^{hi}, CD23^{hi}. The B-1 fraction can be subdivided into B-1a cells that express CD5 (Ly-1), and B1-b cells, which do not. Also, peritoneal B-1 cells are Mac1$^+$, whereas splenic B-1 cells are Mac1$^-$. B-2 B cells do not express either CD5 or Mac1. Mature B-1 cells are maintained by self-renewal of pre-existing cells in the peritoneum, whereas B-2 cells are continually replaced by newly generated cells in the bone marrow. B-1 cells primarily produce antibodies specific for self and bacterial antigens (Ikuta et al., 1992).

Source of cells

Normal mice (e.g. C57/Black6 and BALB/c strains, which are commonly used) are obtained from Harlan, UK or Charles River Suppliers, UK.

Methods of isolation

Materials

Sterile scissors and forceps
Sterile Petri dishes
25-gauge syringe needles
18-gauge syringe needles
2 ml syringes
Universal tubes, 25 ml
Nylon mesh strainers (Falcon 2350)

Solutions and reagents

Phosphate buffered saline (PBS), without Ca^{2+} or Mg^{2+}
Culture medium
Fetal calf serum
Erythrocyte lysis buffer
IL-7
Ficoll–Hypaque (density 1.077 g/ml) (Pharmacia, Uppsala, Sweden)

Preparation of solutions

Bone marrow culture medium

RPMI 1640 with GlutaMAX I™ (Gibco), supplemented with 5% fetal calf serum, 10 ng/ml interleukin 7, 5×10^{-5} M β-mercaptoethanol (BDH), 100 µg/ml gentamicin, 100 U/ml penicillin, 100 µg/ml streptomycin. Gentamicin (Gibco) is supplied as a 10 mg/ml solution. Penicillin and strep-tomycin are prepared as a $100 \times$ stock solution in water, filter sterilised and frozen at $-20\,°C$.

Fetal calf serum

FCS should be batch-tested on unfractionated bone marrow to find a batch that supports both B cell and stromal cell expansion, and does not encourage macrophage outgrowth.

Erythrocyte lysis buffer

1 mM NH_4HCO_3, 114 mM NH_4Cl in distilled water.

Interleukin 7

IL-7 is commercially available in recombinant form (R and D Systems, or Genzyme). Both human and mouse IL-7 activate the murine IL-7 receptor, whereas only human IL-7 will activate the human IL-7 receptor. It is prepared by dissolving the lyophilised powder in PBS/1% FCS (fetal calf serum) at a concentration of 1 µg/100 µl and stored in aliquots at −20°C. It is added to culture medium at a concentration of 10 ng/ml.

Protocols

a. Recovery of lymphocytes from murine bone marrow

1 Remove the femur and tibia of the hind leg and place in a Petri dish. Clean away the attached muscle and fat using scissors and forceps and cut the bones with scissors at either end to expose the bone marrow cavity.
2 Remove the bone marrow by flushing through the cavity with a 25 gauge syringe needle attached to a 2 ml syringe containing culture medium. The marrow plug is removed easily. Scrape the needle around the inside of the bone wall to remove the rest of the cells. (The earliest pre-B cells are found in the peripheral areas of the bone marrow, often embedded in layers of bone-lining cells. Surface IgM^+ cells are distributed in central regions of the marrow.)
3 Vortex the eluate twice for 10 s to break up the marrow plugs and cell clumps to form a single cell suspension and then leave to settle for a minute, allowing small bone spicules to sink to the bottom. Decant the supernatant into a fresh tube. Repeat this procedure until no spicules are evident. This is particularly important if the bone marrow cells are to be sorted as spicules will block the cell sorter.
4 Centrifuge the cell suspension (5 min, 1000 rpm (250g)), 4°C, wash the pellet with PBS and centrifuge again.
5 Resuspend the washed pellet in ice-cold erythrocyte lysis buffer to remove the red blood cells (use 2 ml/mouse), place on ice for 3 min, dilute with ice cold PBS and centrifuge.
6 Resuspend the washed pellet in culture medium containing IL-7 (10 ng/ml) at a concentration of 10^6 cells/ml.

b. Recovery of B lymphocyte precursors from murine fetal liver

1 Remove the liver lobes from 13-to-17-day-old embryos, mince with forceps and push through a nylon mesh strainer.
2 Wash with PBS and lyse RBCs in erythrocyte lysis buffer as above.
3 Resuspend the pellet in culture medium containing IL-7 (10 ng/ml) at a cell density of 10^6/ml.

c. Recovery of B lymphocyte precursors from murine spleen

1 Secure the mouse on its back by its forelimbs. Make an incision in the skin on your right-hand side and cut through the peritoneum immediately underneath to expose the dark red bean-shaped spleen. Free this from connective tissue, mince with forceps and push through a nylon mesh strainer.
2 Wash with PBS and lyse the RBCs in erythrocyte lysis buffer as above.
3 Resuspend the pellet in culture medium at 10^6/ml.

d. Recovery of B lymphocytes from murine peritoneal cavity

1 Make a vertical midline incision through the skin, taking care not to pierce the peritoneal membrane underneath.
2 Pierce the membrane with a 25 gauge needle and inject 10 ml of PBS into the peritoneal cavity. Shake the body gently by holding the hind limbs. Remove the PBS (containing cells) with an 18 gauge needle and syringe.
3 The harvested cells are mostly RBC and lymphocytes. Centrifuge, wash with PBS and lyse the RBC as described above.

e. Recovery of B lymphocytes from human bone marrow

1 Process heparinised bone marrow obtained by iliac crest aspiration as quickly as possible to preserve cell viability.
2 Isolate the mononuclear cells by Ficoll–Hypaque gradient. Dilute the bone marrow sample (1:2) in RPMI with 10% FCS. Layer the diluted sample onto Ficoll in a 2 to 1 ratio.
3 Centrifuge at 1500 rpm ($450g$) for 30 min. Recover the mononuclear cells from the interface between medium and Ficoll. Resuspend the cells in RPMI/10% FCS, wash twice at 1500 rpm for 15 min and resuspend the pellet at 10^6/ml in culture medium with human IL-7.

Establishment and maintenance of cultures

Materials

25 cm² tissue culture flasks (Corning, catalogue number 430168)
24-well tissue culture plates (Nunc, catalogue number 143982)
96-well tissue culture plates (Corning, catalogue number 25860)
Bone marrow culture box

Solutions and reagents

Stromal cell culture medium
Fetal liver stromal cell culture medium
Trypsin/EDTA
5% CuSO$_4$ solution

Preparation of solutions

Stromal cell culture medium is Dulbecco's modified Eagles medium
(DMEM) with GlutaMAX I™ (Gibco), supplemented with 10% FCS,
100 U/ml penicillin and 100 µg/ml streptomycin.
Fetal liver stromal cell culture medium is Iscove's Modified Dulbecco's
medium (IMDM) (Gibco), supplemented with 5% FBS, 10 mM glutamine (Gibco), 100 U/ml penicillin and 100 µg/ml streptomycin.
Trypsin/EDTA is 0.1% trypsin and 0.1% EDTA in PBS.

Preparation of bone marrow culture box

Cut a hole out of the centre of the lid of a large lidded plastic box. Seal this
'window' with clingfilm. Secure a small beaker of 5% CuSO$_4$, and one of
sterile distilled water, inside the box.

Stromal cell cultures:

a. Primary stromal cell cultures from bone marrow

1 Resuspend murine bone marrow at a low cell density (3×10^5 cells/ml)
in 25 cm² flasks in stromal cell culture medium without IL-7 or β-
mercaptoethanol. Incubate in flasks at 37°C, 5% CO$_2$.
2 The next day, shake the flask gently and remove non-adherent cells. The
absence of β-mercaptoethanol and IL-7 will eventually cause lymphocytes
to die.

3 Repeat this procedure every 48 h until a pure population is obtained.

4 The cultures should be 40 to 60% confluent in 3 weeks and can be used for sub-culture of the proliferating B lymphocyte cultures. Because they grow slowly, they do not need to be irradiated when cultured with B cell precursors. Confluent cultures can be replated at 10^6 cells per T175 flask.

5 Two days before use in bone marrow cultures, remove spent medium and replace with fresh medium (to condition it with important secreted factors).

6 A similar procedure is used to derive stromal cell lines from human bone marrow, using the purified mononuclear cell fraction plated at low cell density as above.

b. Primary stromal cell cultures from murine fetal liver

1 Resuspend fetal liver cells at a density of $5-10 \times 10^6$ cells in fetal liver stromal cell culture medium. Incubate in 25 cm² flasks at 37 °C, 5% CO_2.

2 Remove adherent cells as above (steps 2 and 3).

3 Fetal liver stromal layers often grow faster than bone-marrow-derived stroma and thus may need to be gamma irradiated (20 Gy, equals 2000 rad) before the addition of B cells.

c. Stromal cell lines

A number of well-characterised stromal cell lines are available that support B lymphopoiesis *in vitro*. These include ST2, which secretes IL-7, and PA6 and S17, which do not. These lines are especially useful for the culture of sorted populations of lymphocytes, from which primary stromal cells have been removed.

1 Grow stromal lines in stromal cell culture medium to no greater than 80% confluence.

2 Subculture by removing medium, rinsing flask with EDTA and adding trypsin/EDTA for a short time, until the cells detach with gentle tapping. Add stromal cell culture medium to neutralise the trypsin, and pellet cells by centrifugation at 1000 rpm (250g) for 5 min at 4 °C.

3 Place in culture with bone marrow cells at a confluence no greater than 30%. Higher confluence encourages the outgrowth of macrophages, which will overrun the culture and inhibit B lymphopoiesis.

4 As stromal cell lines grow faster than primary stromal layers, they must be irradiated (20 Gy) to prevent overgrowth.

5 In our experience PA6, which is a pre-adipocyte cell line, must be maintained at no more than 70–80% confluence. If allowed to overgrow, the cells terminally differentiate into adipocytes containing microscopically visible fat bodies.

Long-term murine bone marrow cultures

The RBC-lysed bone marrow suspension from a C57/BL6 mouse contains approximately 50% macrophages, 30% B lymphocytes and small numbers of T cells and natural killer cells.

1 On the day before collection of the bone marrow, γ-irradiate a stromal cell line such as ST2 or PA6 at 20 Gy. Plate cells out in 25 cm^2 (T25) flasks or 24- or 96-well plates in stromal cell culture medium at a confluence no greater than 30%. (Take into account that the cells will probably continue to divide for a short period after irradiation.) Alternatively, use a primary stromal cell layer prepared as described above.

2 The next day, add the bone marrow suspension at a (lymphocyte) cell density of 10^6 per ml of bone marrow culture medium – 5 ml of suspension are cultured per T25 flask, 1 ml/well in 24-well plates, and 200 μl/well in 96-well plates.

3 Gas the flasks gently with CO_2, close the lids tightly and incubate at 37 °C.

4 As cultures can grow for months, maintaining sterility can be problematic if plates are used. Enclose these in a bone marrow culture box and incubate at 37 °C, 5% CO_2. This allows the cultures to remain equilibrated at 5% CO_2, prevents dehydration and protects from airborne contamination.

5 Feed the cultures once or twice weekly by carefully aspirating 50% of the spent culture medium, taking care not to disturb the non-adherent cells. Replace with fresh bone marrow culture medium.

6 To harvest lymphocytes for subculture or analysis, pipette the medium up and down gently. The loosely attached B precursors will detach from the stromal cells, most of which should remain firmly attached. If the few contaminating stromal cells in the suspension must be removed, transfer it to a fresh flask for 4 h at 37 °C. The stromal cells will attach to the plastic and the lymphocytes can be harvested by pipetting.

7 To subculture lymphocytes, once they have expanded after 2–3 weeks, transfer the suspension of detached cells to a new flask containing a stromal cell line or a primary stromal culture. The original culture can be perpetuated by leaving 20% of the suspension in the flask, and replenishing with fresh medium.

8 Long-term cultures derived from fetal liver are established as described above.

9 Once cultures are well established, DMEM can be used instead of RPMI to achieve higher growth rates. It is important to use only RPMI at early stages as it decreases fibroblast outgrowth.

Cultures of early pro-B cells from bone marrow

Witte and co-workers have described two methods developed to selectively grow B lineage restricted progenitors (early pro–B cells) from murine bone marrow. These are the B220$^+$, stromal-dependent lymphocytes, which do not yet require c-kit ligand or IL-7 for growth, retain germline Ig configuration, but have the potential to develop into pre-B cells. In the earlier method, bone marrow cells were retrovirally transduced with the BCR/ABL gene and plated onto the stromal cell line, S17. A subsequent method selectively amplifies pro–B cells on an S17 stromal layer without requiring BCR/ABL transduction (Faust et al., 1993).

Fetal liver lymphocyte cultures

These can be established essentially as described for bone marrow cultures. A stromal cell layer is required. IL-7 is not required for differentiation of B220$^-$ fetal liver cells to the B220$^+$ stage, but is essential for later stages in development.

Description of the long-term murine bone marrow culture

Under the conditions described above, the more immature B lymphocyte fractions adhere to the stromal cell line and/or the primary stromal cells within 24 h. Those that have rearranged their IgH chain remain in suspension. During the next 7–14 days, the macrophages and T cells will gradually die due to lack of appropriate growth factors, i.e. IL-3, GM-CSF, IL-2, etc. The stromal cells will expand slowly. The more mature B cells do not expand significantly. Between approximately day 7 and day 14, non-adherent cells will be very sparse and the cultures may appear to reach a 'crisis' as described by Whitlock and Witte, in which there appear to be few surviving or proliferating B cells. After a few days, a small number of colonies of small, round, adherent pro-B lymphocytes will appear and should expand rapidly. As they rearrange IgH genes and acquire independence from stromal cell contact, they detach and appear as small clusters 'like bunches of grapes' or as single cells in suspension.

Longer-term culture and passage

Stromal cell cultures and lines should be maintained at a confluence no greater than 80% and should be passaged after detachment using trypsin/EDTA. Stocks of early passages should be frozen. Cells should be sub-cultured twice weekly and a fresh batch used after 10 to 12 passages. Bone marrow cultures should be propagated by periodic transfer to a fresh stromal feeder layer.

Cryopreservation

Solutions and reagents

Freezing medium is 90% FCS with 10% dimethylsulphoxide (DMSO).
Bone marrow freezing medium is 50% bone marrow culture medium, 40% FCS and 10% DMSO.

Protocol

Stromal cell cultures and lines: Cells are removed from flasks, pelleted by centrifugation and resuspended on ice in 1 ml aliquots of cold freezing medium. These are cooled slowly overnight by transfer in a polystyrene container to $-20\,^\circ$C and then to $-70\,^\circ$C. They are subsequently stored in liquid nitrogen.

Primary bone marrow cultures: As above, but for the use of a different medium (bone marrow freezing medium).

Cells should be thawed out by warming rapidly in a $37\,^\circ$C waterbath, followed by centrifugation to remove the freezing medium. Thawed cells are resuspended in the appropriate culture medium, warmed to $37\,^\circ$C.

Applications: analysis of developing lymphocytes in culture

Immunofluorescence analysis and cell sorting

The cell surface and cytoplasmic markers that characterise stages in B cell development can be detected with fluorescent antibodies, either on slide preparations or using fluorescence activated cell sorter (FACS) analysis. Antibodies can be chemically linked to any one of a variety of fluorochromes – fluorescent dye molecules whose key property is that they emit light at different wavelengths. Optical filters and electronic detectors convert this light to electronic signals for FACS analysis. Common fluorochromes used in such experiments are fluorescein isothiocyanate (FITC), ʀ-Phycoerythrin (PE) and Red-670. FITC emits green light at 514 nm, PE emits orange–red light at 600 nm and Red-670 emits red light at 670 nm. Stained cells are passed through the FACS analyser, which detects and displays them graphically according to which fluorochromes have bound. This information can be used for analysis. Distinct cell populations can also be physically separated and recovered.

Antibodies are now available individually conjugated to a number of different dyes with distinct emission spectra. With most machines, it is possible to stain cell populations with three conjugated antibodies simultaneously, provided the emission spectra of the conjugates are sufficiently distant. For example: IgM-FITC with B220-PE and CD43-Red-670 are commonly used for the analysis of B lymphopoiesis in the bone marrow. Provided the correct lasers are available, four-colour analyses are also possible.

Materials

6 ml round-bottomed polystyrene FACS tubes (Becton Dickinson, catalogue number 2058)
96-well tissue culture plates (Corning, catalogue number 25860)

Solutions and reagents

FACS buffer is PBS, supplemented with 1% FCS
Normal goat IgG (Sigma, catalogue number I-5256)
FC Block™ (Pharmingen), which is a rat anti-mouse CD16/CD32 monoclonal antibody that blocks the FCRγII (CD32) and FCRγIII (CD16) low affinity Fc receptor binding sites to prevent non-specific staining
FACS lysing solution™ (Becton Dickinson). This solution simultaneously lyses erythrocytes and permeabilises white cells
Propidium iodide (Sigma)
4% formaldehyde, dissolved in PBS
0.2% Triton X100, dissolved in PBS

Antibodies for murine B cell analysis

Monoclonal antibodies against murine B220, CD19, CD43, CD117 (c-kit) CD24 (HSA), CD25, CD5, CD22 and CD23 are available from

Pharmingen (San Diego, CA). Polyclonal goat anti-mouse IgM and goat anti-mouse kappa are available from Southern Biotech (Birmingham, AL).

Antibodies for human B cell analysis

CD34, CD38 (Becton Dickinson, Mountain View, CA); CD10, CD19 (Dakopatts, Glostrup, Denmark); CD25 (Pharmingen, San Diego, CA); CD24 (Boehringer Mannheim GmbH, Germany); TdT (Immunotech). Polyclonal antibodies (F(ab)2 fragments of rabbit antibodies) against human IgM and kappa are supplied by Dakopatts.

Protocols for FACS analysis of B lymphocyte populations in culture

a. Cell surface staining for FACS analysis

Depending on sample numbers, this can be carried out directly in FACS tubes, or in 96-well plates.

1 Resuspend cells in 50 μl FACS buffer at 10^5 to 10^6 per sample.
2 Add 50 μl of 2×concentrated antibody solution. This is made up in FACS buffer containing an excess of, for example, normal goat IgG, to prevent non-specific binding. Alternatively, the cells can be preincubated with FC Block™.
3 Incubate 30 min to 1 h on ice in the dark.
4 Fill the tube up with FACS buffer and centrifuge to wash the cells (5 min, 1000 rpm, 4 °C). For directly conjugated antibodies, wash the cells again and resuspend in 1 ml of FACS buffer for analysis. If using 96-well plates, add 100 μl of FACS buffer, spin at 1000 rpm for 1 min, invert the plate rapidly to remove the medium and wash again.
5 For antibodies not directly conjugated to a fluorescent dye, e.g. biotin-ylated antibodies, resuspend in 50 μl of FACS buffer and add 50 μl of secondary reagent (e.g. streptavidin–FITC).
6 Incubate 15 to 30 min on ice, then wash twice by centrifugation and resuspend in 1 ml of FACS buffer for analysis.
7 Propidium iodide (PI): In non-permeabilised samples, this dye is taken up by dead cells and emits fluorescently when intercalated with DNA. It is thus useful for identification of dead cells that can then be excluded from the analysis. After the first wash in step 6, add 100 μl PI, fill the tube up with PBS, centrifuge, and resuspend pellet as before. PI staining can be analysed in either the FL2 or FL3 channel on the FACS analyser.

b. Intracellular staining

(1) Cytoplasmic staining:

1 Fixation: Resuspend cell pellet in 200 µl 4% formaldehyde (in PBS) (4% paraformaldehyde gives less background) for 5 min on ice.
2 Permeabilisation: Without centrifugation, add 200 µl of 0.2% (v/v) Triton X100 (in PBS) for 5 min on ice.
3 Dilute to 4 ml with FACS buffer, centrifuge, and resuspend pellet in FACS buffer.
4 Alternatively to steps (1)–(3) use FACS lysing solution™. Subsequent staining and washing procedures are performed as described for surface-stained cells.

(2) Nuclear staining: The following method allows combined surface and nuclear staining. It can be used on small numbers of cells, such as sorted populations. Thus, expression of nuclear antigens such as human terminal deoxynucleotide transferase (TdT) can be analysed.

1 Fix the cells in 1 ml of ice-cold 1% paraformaldehyde in PBS for 2 min on ice.
2 Without centrifuging, add 1.5 ml of ice-cold absolute methanol and incubate on ice for 20 min.
3 Wash the cells twice with ice-cold FACS buffer (800g, 10 min, 4 °C).
4 Resuspend the cells in an appropriate volume of fetal calf serum to block for 15 min.
d Without centrifuging, add a mixture of three FITC-conjugated anti-human TdT monoclonal antibodies (Immunotech) at a final concentration of 1:20 for 30 min on ice.

Subsequent washing procedures and FACS analysis are carried out as described for surface-stained cells. With this method, three-colour analysis can be carried out using PE and, for example, Red-670 conjugated surface staining antibodies.

c. FACS Sorting

The staining and washing procedures are identical to those used for FACS analysis, but on a larger scale. Cell numbers for bone marrow sorting are between 10^7 and 10^8. Thus staining is carried out in a total volume of 2–4 ml, using the same concentration of antibodies as for FACS analysis. All staining and washing procedures are carried out on ice using ice-cold solutions. Cells are resuspended in FACS buffer at a concentration of $5–10 \times 10^6$ per ml for sorting and are collected into 1 ml of culture medium with 20% FCS. If cells are being sorted for subsequent cell culture, all procedures are carried out under sterile conditions and as quickly as possible to maintain cell viability.

d. FACS analysis of cell cycle progression and apoptosis

Solutions and reagents

Phosphate buffered saline (PBS)
PBS/0.5% (v/v) Tween 20
70% ethanol
RNase (1 μg/ml) (Sigma)
Propidium iodide (400 μg/ml) (Calbiochem)
5-bromo-2'-deoxyuridine (BrdU) (Sigma)
Anti BrdU antibody, FITC-conjugated (Pharmingen)
DNA denaturing solution (3M HCl/0.5% Tween 20 in water)
Sodium tetraborate solution (0.01 M in water)
Apoptosis Detection Kit (R and D Systems, UK)

(1) Measurement of DNA content To determine the proportion of cells in each stage of the cell cycle:

1 Resuspend cells in 200 μl PBS.
2 Add 2 ml of ice cold 70% ethanol and incubate for 30 min on ice.
3 Centrifuge and resuspend pellet in 800 μl PBS. If the cells are clumped, pass through a 25 gauge syringe needle.
4 Add 100 μl RNase and 100 μl propidium iodide and incubate at 37°C for 30 min
5 Wash twice with PBS, resuspend in PBS and analyse at 488 nm. Gate to exclude cell doublets. Measurement of number of cells versus DNA content (PI staining) determines the number of cells at each stage of the cell cycle. A typical plot of an asynchronous cycling cell population will contain a sharp diploid (2N) peak at the G_0/G_1 stage, followed by a broader peak representing the S phase, and a second peak (4N) for the G_2/M phase. Cells with a sub-diploid DNA content are usually apoptotic (Fig. 5.3B).

A

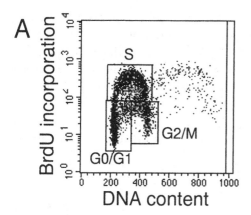

B

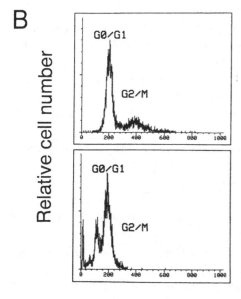

DNA content

Fig. 5.3. Cell cycle analysis. (A) Staining with an anti-BrdU antibody is plotted on the y-axis, against DNA content by PI staining, on the x. Boxes mark different cell cycle phases. (B) PI staining on the x-axis is plotted against relative cell number. Cells with sub-diploid DNA content in the lower panel are apoptotic (modified from Corcoran et al., 1996).

(2) BrdU labelling and cell cycle analysis The thymidine analogue 5-bromo-2'-deoxyuridine (BrdU) is widely used to study lymphocyte lifespans and population dynamics. FACS analysis of BrdU incorporation into cells, when plotted against the total DNA content measured by PI staining, indicates the proportions of cells that are in G_0/G_1, S or G_2/M stages of the cell cycle (Fig. 5.3A).

1 *In vitro* labelling: Pulse lymphocyte cultures with 10 μM BrdU for varying times.
2 *In vivo* labelling: Administer BrdU (1 mg/ml) to mice in drinking water for varying times.
3 Analysis of BrdU incorporation: Harvest cells from culture or recover from mice, sort if necessary, and wash with PBS.
4 Fix 10^5 cells in 500 μl 70% ethanol (stored at −20 °C, added drop-wise, mixed well), for 20 min at room temperature (RT).
5 Wash cells twice in PBS and incubate with 500 μl of DNA denaturing solution for 20 min at RT to denature the DNA. Wash once with PBS.
6 Resuspend the pellet in 250 μl sodium tetraborate solution for 5 min at RT. Wash twice with PBS/0.5% Tween 20.
7 Add 20 μl of FITC-conjugated anti BrdU antibody and incubate at RT for 30 min.
8 To double stain with PI if required, add 10 μl PI (4 mg/ml) and incubate at RT for 10 min.
9 Wash and resuspend in FACS buffer.

(3) Annexin V labelling A useful early marker of apoptosis is binding of Annexin V, a high-affinity phosphatidylserine binding protein found largely at the cytosolic face of plasma membranes. During apoptosis, cells lose their cell membrane phospholipid asymmetry and expose phosphoserine on the outer membrane. Differential binding of fluorochrome-labelled Annexin V and PI can be used to identify cells undergoing apoptosis and distinguish them from necrotic cells. The necessary reagents are available in kit form from a number of companies.

Preparation of DNA and RNA for analysis by PCR

Analysis of B cell development is often limited by the small numbers of cells that can be isolated from various fractions. Therefore, an important technique for the detailed analysis of gene expression in fractions of developing B cells, and of their immunoglobulin rearrangement status, is the polymerase chain reaction (PCR). PCR analyses can be carried out on single cells or small sorted populations.

Preparation of genomic DNA for the analysis of Ig rearrangements by PCR

This methodology was originally developed by Ehlich *et al.* (1994) and extended by ten Boekel *et al.* (1995). Modifications to these methods have been incorporated by us (Corcoran *et al.*, 1998). Nested primers and two rounds of PCR amplification are used to detect individual (D)J, V(D)J and V kappa rearrangements. The protocol below describes the preparation of genomic DNA from bone marrow cells.

Materials

96-well polycarbonate plates appropriate for the PCR thermal cycler.

Solutions and reagents

PCR mix: 30 mM Tris HCl, pH 8.3, 150 mM KCl, 4.5 mM MgCl$_2$, 0.003% gelatine, 0.15 mM 2-mercaptoethanol, 1 μg/ml tRNA

DNA isolation buffer: 10 mM Tris HCl, pH 8.3, 50 mM KCl, 2 mM MgCl$_2$, 0.45% Tween-20, 0.45% NP40

Proteinase K (Sigma)

a. Isolation of genomic DNA from bone marrow cells

(1) Single cell method:

1 Sort single cells into 96-well polycarbonate plates containing 10 μl of PCR mix. This can be done with an automatic cell deposition unit on the FACS sorter. If this is not available, sorted populations can be diluted by hand into 96-well plates. Spin the plates at 1000 rpm (250*g*) for 30 sec to ensure that the cell has been deposited in the buffer.

2 Add Proteinase K at 60 μg/ml, overlay the samples with mineral oil and digest at 55 °C for 1 h.

3 Inactivate Proteinase K for 5 mins at 95 °C. The plates can then be stored at −70 °C.

(2) For bulk populations:

1 Spin down cells, wash the pellet with PBS and centrifuge again.
2 Resuspend the pellet in DNA isolation buffer at a concentration of 5×10^5/ml (i.e. 100 000 sorted cells in 200 μl).
3 Add Proteinase K at 60 μg/ml and incubate at 55 °C for 1 h.
4 Inactivate the Proteinase K for 5 min at 95 °C.
5 Store the genomic DNA at 4 °C.

Details of the primers and conditions for PCR reactions are provided in the references cited above.

Preparation of cDNA for RT-PCR analysis of gene expression

The transcription of many of the markers that identify different stages in B cell development can be measured by reverse transcription, followed by PCR using specific primers (RT-PCR). This can be done for both single cells and bulk or sorted populations. Characteristic genes whose expression can distinguish progressive stages of development in this way include (Li *et al.*, 1993): *RAG1, RAG2, TdT, Igα, Igβ, λ5, VpreB, BP-1*. Another important application of RT-PCR assays is in the detection of the sterile transcription of unrearranged gene segments from specific regions of the Ig loci (Corcoran, *et al.*, 1998). Sterile transcription is correlated with the chromatin changes that precede V(D)J rearrangement.

Materials

RNAzol B™ (Biogenesis, Poole, Dorset, England)
Super RT reverse transcriptase (HT Biotechnology, Cambridge, England)
RNasin (Promega, Madison, WI, USA), which is an RNase inhibitor

1 Preparation of RNA: When studying sorted B cell populations the number of cells obtained is often rate-limiting. Sort single cells in PBS. Heat to 65 °C for 2 min, put on ice and perform a 10 μl reverse transcription (RT) reaction as described below. For bulk populations of 20 000 to 100 000 cells we use a scaled-down version (100 μl RNAzol B) of the RNAzol B™ method, including 10 to 30 μg of transfer RNA as a carrier.
2 Reverse transcription: We use 'Super RT' reverse transcriptase according to the manufacturer's instructions. It is important to include an RNase inhibitor such as RNasin.

Details of the primers and conditions for PCR reactions are provided in the references cited above.

Genetic manipulation of developing B lymphocytes

Transgenesis

The use of transgenic mice in which genes of importance to B lymphocyte development have been introduced or, alternatively, disrupted by gene targeting, has been a crucial advance in the study of B lymphopoiesis. Space here does not permit an extensive review of this subject, and the reader is referred instead to excellent reviews (e.g. Loffert *et al.*, 1994). Suffice it to mention that transgenic expression of Ig genes has enabled the discovery and molecular dissection of fundamental processes including allelic exclusion and somatic hypermutation, whilst gene targeting has been used to explore the mechanisms of V(D)J recombination, the functions of surface receptors and their ligands involved in the development and activation of B lymphocytes, and the transcription factors that dictate cell fate during B lymphopoiesis. Needless to say, this list is by no means exhaustive.

Retrovirus-mediated gene transfer

A significant drawback in using transgenesis is that it is laborious and time-consuming. A rapid alternative for the genetic manipulation of developing B lymphocytes is retrovirus-mediated gene transfer. When used in cultures of bone marrow derived from mice harbouring targeted gene disruptions, this method provides a particularly powerful tool for the functional analysis of molecules encoded by the targeted genes.

Retroviruses are suited as vehicles for gene delivery into mammalian cells because their life cycle includes a stage in which the viral genome is integrated into that of the host. Once integration is achieved, endogenous retroviral regulatory sequences direct the expression of virally encoded genes at a fairly high level. Recently developed retroviral gene delivery systems include two basic components: a retroviral vector, in which the gene of interest can be cloned; and a packaging cell line, in which infectious virions carrying the gene of interest can be produced.

Retroviral vectors consist, at their simplest, of a backbone with the 5′ and 3′ long terminal repeat (LTR) regulatory elements flanking the ψ packaging signal and a polylinker to facilitate cloning of a gene of interest (Fig. 5.4). Retroviral *gag*, *pol* and *env* genes encoding functions required for replication and assembly of infectious virions are deleted in the vector. Instead, these

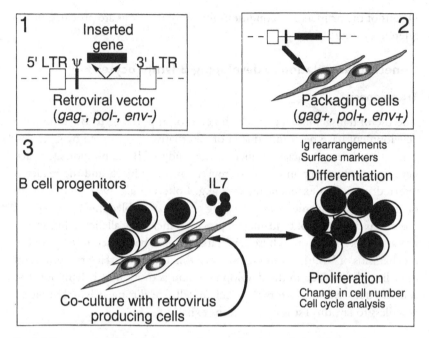

Fig. 5.4. Retrovirus mediated gene transfer into developing B lymphocytes. Steps 1–3 are shown as described in the text.

functions are provided in *trans* by a packaging cell line (Fig. 5.4), with the host range of the resulting virion determined by the *env* function. Ecotropic *env* genes permit infection of murine cells, whilst the host range of amphotropic *env* is extended to include human cells, necessitating special precautions for biological containment. Recombinant virions produced in this way can infect cells only once, since the helper functions required for further replication are not carried within the viral genome.

We use retroviral vectors based upon the Moloney leukaemia virus genome for gene delivery (Corcoran *et al.*, 1996), the features of which are similar to the MFG vector developed by Mulligan and co-workers (Riviere, Brose & Mulligan, 1995). An overview of the protocol used for gene delivery into murine B lymphocyte precursors is provided here, because space does not permit a fuller discussion. Briefly, ecotropic packaging cell lines are transfected with 15–20 µg of supercoiled vector DNA mixed in a 10:1 ratio with a plasmid encoding neomycin resistance (e.g. pWCneo, Stratagene) by the calcium phosphate method. A kit for transfection (Stratagene) is used according to manufacturer's recommendations. Stable transfectants are selected in 1 mg/ml G418 sulphate (Gibco-BRL), and multiple vials cryo-preserved at an early (<4) passage in culture.

For retroviral transduction, stable retroviral producer cells are washed free of selective medium, plated to 60% confluence in a 75 cm^2 flask and sub-jected to 2000 rad γ-radiation (20 Gy). After incubation of the producer cells for 18–24 h at 34 °C in 10 ml IL-7 containing growth medium described previously, fractions of bone marrow B cell progenitors isolated by FACS sorting (or unfractionated bone marrow depleted of plastic-adherent cells) are added at 0.5–1 × 10^6 cells/ml in a minimal volume of the same medium. Co-culture is continued for 3–5 days at 37 °C under conditions described for bone marrow cultures, by which time expression of retrovirally encoded genes is generally detectable in 50–70% of lymphoid progenitors.

A critical parameter for optimal transduction is the titre of infectious virions released by the stable packaging cell line. High-titre producers should routinely be isolated by screening for efficient expression of the retrovirally encoded gene of interest. Isolation by FACS sorting is possible when the gene of interest encodes a surface protein. Alternatively, dilution subcloning permits screening for the expression of intracellularly localised proteins encoded by the gene of interest. Newer vector designs incorporate internal ribosome entry sequences 3′ of the inserted gene of interest to link expression with that of a reporter such as green fluorescent protein encoded within the same polycistron.

Culture of mature B lymphocytes

The techniques thus far discussed pertain to cultures of developing B lymphocytes from the bone marrow. However, many important steps in B lymphocyte development and activation – including, for example, somatic hypermutation and class switching – occur in peripheral lymphoid tissues and involve sIgM$^+$sIgD$^+$ or subsequent maturational stages. Reliable techniques for the long-term culture of such cells without transformation have not yet been developed. Several advances have been made, although a detailed description is beyond the scope of this chapter. Transformed cell lines have been identified that undergo processes such as somatic hyper-mutation *in vitro*. Although it is as yet unclear if the molecular basis of *in vitro* hypermutation is the same as that which occurs *in vivo*, many of its charac-teristics are identical, making it an attractive model system for analysis. An *in vitro* system has been developed for the study of somatic hypermutation in which primary peripheral B cells are cultured in the presence of dendritic cells, cytokines and CD40L. Under these conditions, there is little long-term proliferation, but cells survive for extended periods sufficient to detect somatic hypermutation *in vitro*. Finally, it should be noted that techniques for the creation of human lymphoblastoid continuous lines following immortalisation with the Epstein–Barr virus are widely used. These cells retain some (but not all) characteristics of primary B cells, potentially facilitating the study of processes such as BCR signalling.

Summary

Methods for murine B lymphocyte culture originally defined by Whitlock and Witte have since been refined by characterisation of the stromal elements, contact-dependent signals and cytokines that promote B lymphopoiesis. In conjunction with techniques of gene targeting, and more recently retroviral gene transfer, to genetically manipulate murine B cell progenitors, these culture methods have been responsible for the rapid increase in understanding of the molecular and cell biology of lymphopoiesis. The development and application of similar culture methods to human B cell progenitors promises to be of great value in the study and, eventually, therapy, of immune deficiencies and lymphoid malignancies.

References

Corcoran, A. E., Smart, F. M., Cowling, R. J., Crompton, T., Owen, M. & Venkitaraman, A. R. (1996). The interleukin-7 receptor a chain transmits distinct signals for proliferation and differentiation during B lymphopoiesis. *EMBO J.*, **15**, 1924–32.

Corcoran, A. E., Riddell, A., Krooshoop, D. & Venkitaraman, A. R. (1998). Impaired immunoglobulin gene rearrangement in mice lacking the interleukin-7 receptor. *Nature*, **391**, 904–8.

Ehlich, A., Martin, V., Muller, W. & Rajewsky, K. (1994). Analysis of the B-cell progenitor compartment at the level of single cells. *Curr. Biol.* **4**, 573–84.

Faust, E. A., Saffran, D. C., Toksoz, D., Williams, D. A. & Witte, O. N. (1993). Distinct growth requirements and gene expression patterns distinguish progenitor B cells from pre-B cells. *J. Exp. Med.*, **177**, 915–23.

Ghia, P., Gratwohl, A., Signer, E., Winkler, T. H., Melchers, F. & Rolink, A. G. (1995). Immature B cells from human and mouse bone marrow can change their surface light chain expression. *Eur. J. Immunol.*, **25**, 3108–14.

Ghia, P., ten Boekel, E., Sanz, E., de la Hera, A., Rolink, A. & Melchers, F. (1996). Ordering of human bone marrow B lymphocyte precurors by single-cell polymerase chain reaction analyses of the rearrangement status of the immunoglobulin H and L chain gene loci. *J. Exp. Med.*, **184**, 2217–29.

Hardy, R. R., Carmack, C. E., Shinton, S. A., Kemp, J. D. & Hayakawa, K. (1991). Resolution and characterisation of pro-B and pre-pro-B cell stages in normal mouse bone marrow. *J. Exp. Med.*, **173**, 1213–25.

Hayashi, S., Kunisada, T., Ogawa, M., Sudo, T., Kodama, H., Suda, T., Nishikawa, S. & Nishikawa, S. (1990). Stepwise progression of B lineage differentiation supported by interleukin 7 and other stromal cell molecules. *J. Exp. Med.*, **171**, 1683–5.

Ikuta, K., Uchida, N., Friedman, J. Weissman, I. L. (1992). Lymphocyte development from stem cells. *Annu. Rev. Immunol.*, **10**, 759–81.

Li, Y. S., Hayakawa, K. & Hardy, R. R. (1993). The regulated expression of B lineage associated genes during B cell differentiation in bone marrow and fetal liver. *J. Exp. Med.*, **178**, 951–60.

Loffert, D., Schaal, S., Ehlich, A., Hardy, R. R., Zou, Y. R., Muller, W. & Rajewsky, K. (1994). Early B-cell development in the mouse: insights from mutations introduced by gene targeting. *Immunol. Rev.*, **137**, 135–53.

Melchers, F., Haasner, D., Grawunder, U., Kalberer, C., Karasuyama, H., Winkler, T. & Rolink, A. (1994). Role of Ig H and L chains of surrogate L chains in the development of cells of the B lymphocyte lineage. *ARI*, **12**, 209–25.

Nemerow, G. R., McNaughton, M. & Cooper, M. D. (1985). Binding of monoclonal antibody to the Epstein Barr virus/CR2 receptor induces activation and differentiation of human B lymphocytes. *J. Immunol.*, **135**, 3068–73.

Riviere, I., Brose, K. & Mulligan, R. C. (1995). Effects of retroviral vector design on expression of human adenosine deaminase in murine bone marrow transplant recipients engrafted with genetically modified cells. *Proc. Natl. Acad. Sci. USA*, **92**, 6733–7.

Rolink, A., Haasner, D., Nishikawa, S. & Melchers, F. (1993). Changes in the frequencies of clonable preB cells during life in different lymphoid organs of mice. *Blood*, **81**, 2290–5.

Rolink, A., ten Boekel, E., Melchers, F., Fearon, D. T., Krop, I. & Andersson, J. (1996). A subpopulation of B220+ cells in murine bone marrow does not express CD19 and contains natural killer cell progenitors. *J. Exp. Med.*, **183**, 187–194.

Schatz, D. G., Oettinger, M. A. & Schlissel, M. S. (1992). V(D)J recombination: molecular biology and regulation. *ARI*, **10**, 359–83.

ten Boekel, E., Melchers, F. & Rolink, A. (1995). The status of Ig loci rearrangements in single cells from different stages of B cell development. *Int. Immunol.*, **7**, 1013–19.

Whitlock, C. A. & Witte, O. N. (1987). Long-term culture of murine bone marrow precursors of B lymphocytes. *Methods Enzymol.*, **150**, 275–91.

Yasunaga, M., Wang, F.-H., Kunisada, T., Nishikawa, S. & Nishikawa, S.-I. (1995). Cell cycle control of c-kit+IL7R+ precursor B cells by two distinct signals derived from IL7 receptor and c-kit in a fully defined medium. *J. Exp. Med.*, **182**, 315–20.

6

Monocytes and macrophages

James A. Mahoney, Richard Haworth and Siamon Gordon

Introduction

Monocytes and macrophages (Mϕ) are critical players in both natural and acquired immune responses. Mϕ exist in virtually every tissue in the body, and show striking morphological heterogeneity in different tissues. These differing morphologies are a demonstration that Mϕ can 'tune' their gene expression and functions to suit the requirements for a particular site. Monocytes are the blood-borne precursors of Mϕ. In the event of infection, tissue damage or other injury, large numbers of monocytes are recruited from the bloodstream into the site of the insult and differentiate into the appropriate Mϕ phenotype.

The primary role of Mϕ in immunity is as a professional phagocyte. Phagocytosis by Mϕ can be through either natural (e.g. via mannose receptor and scavenger receptor) or acquired (via Fc receptor and complement receptor) immune mechanisms (Gordon, 1996). Mϕ are also important modulators of a variety of immune responses, via secretion of a large array of cytokines and chemokines. Moreover, Mϕ have numerous roles outside the immune system, with functions in such diverse areas as haematopoietic development, tissue homeostasis and wound repair. Examples include phagocytosis of apoptotic thymocytes in the thymus, putative cell–cell interactions in the bone marrow, and stimulation of angiogenesis at the sites of tissue injury.

The majority of the steps in the monocytic differentiation pathway occur in the bone marrow (reviewed by van Furth, 1992). Bipotential granulocyte–monocyte stem cells give rise to monoblasts. A monoblast divides to become two promonocytes, each of which divides again to become two monocytes. The monocytes then leave the bone marrow and enter the circulation, from which they will be recruited into the tissue and differentiate into Mϕ.

Table 6.1. *Classes of macrophage activation*

	Activation	Alternative activation	Inactivation
T helper type	Th1	Th2	Th2
Cytokines	IFNγ	IL-4, IL-13	IL-10
MHC II	↑↑	↑	↔/↓
iNOS	↑↑	↓	↓↓
Legionella growth	↓	↔	↑
Mannose receptor	↓↓	↑↑	↑
IL-1 RA	↑	↑↑	↑
Growth	↓	↑	↔

Note:
Adapted from Gordon *et al.*, 1995; Park & Skerrett, 1996 and personal observations.

Mature Mφ exist in a number of different activation states (Table 6.1). Stromal and other tissue-resident Mφ populations are normally at rest. In response to interferon γ (IFNγ) secreted by Th1 cells and natural killer cells, Mφ (either resident or recruited from the monocyte pool) become activated for cell killing (Fig. 6.1). The activation programme is potentiated by Mφ production of interleukin (IL)-12 and IL-18, which leads to further IFNγ release. In contrast, an IL-4 or IL-13 signal from Th2 cells results in induction of an alternative activation pathway, wherein mechanisms for antigen presentation and phagocytosis of certain particles are upregulated. IL-10, secreted by both Mφ and Th2 cells, can block progression towards either of these forms of activation, and thus has been termed an inactivation signal. However, this view may be an oversimplification, since many genes are induced after IL-10 treatment (personal observations).

Dendritic cells, the cells primarily responsible for presenting antigens to T cells, are closely related to Mφ. Like Mφ, dendritic cells can be produced *in vitro* from blood monocytes. Both cell types share the ability to present antigen and to phagocytose particles. However, dendritic cells present antigen to T cells to a much greater extent than Mφ, while Mφ are much more actively phagocytic than dendritic cells. Techniques for culturing dendritic cells are discussed in Chapter 2.

The vast majority of research into Mφ function has been conducted using either human or murine cells, and a variety of Mφ-specific markers have been identified and characterised in both species. Clinical relevance aside, the primary advantage of working with human Mφ is that large numbers of monocytes can be obtained from human blood and subsequently allowed to

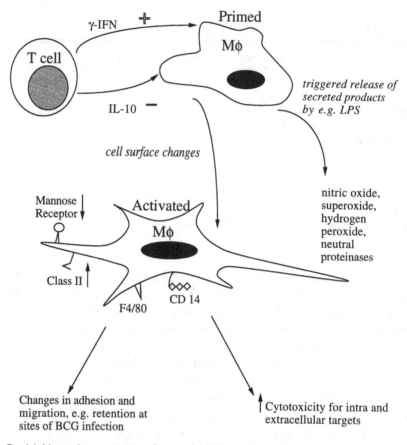

Fig. 6.1. Macrophage activation. Priming by IFNγ results in increased release of effector molecules and changes in cell surface molecules, e.g. class II MHC and the mannose receptor. See text for further details.

differentiate into Mφ in culture. Meanwhile, working with the mouse has the advantage that mature populations of tissue–specific Mφ can often be isolated and studied directly from tissues not normally available from humans.

Several cell lines displaying characteristics of the monocytic lineage are available in both the mouse and the human. Because these cell lines grow easily and rapidly, they can be of great value for obtaining large quantities of Mφ-restricted molecules of interest, and for conducting certain assays of Mφ function. However, these cell lines tend to display a phenotype most like the early cell types in the monocytic lineage, monoblasts and promonocytes.

This is not surprising, since monocytes and Mϕ generally do not divide. While some of the hallmarks of monocytes and Mϕ are present in these lines, many others are not. For example, cell surface markers of Mϕ, such as Mϕ mannose receptor and Fc receptors, are not expressed or weakly expressed in most monocytic cell lines. As a result, isolation and culture of primary monocytes/Mϕ is frequently the most appropriate method for studies of Mϕ function.

In the first section of this chapter, we will describe methods for isolation of human monocytes and culture of human monocyte-derived Mϕ. The next section will describe the isolation and culture of murine Mϕ from a variety of tissues. A few of the many potential applications of these primary cell cultures will be discussed.

Primary human macrophage culture

Human Mϕ are most frequently obtained by purification of circulating blood monocytes, followed by *in vitro* differentiation of monocytes into Mϕ. The protocol involves separation of peripheral blood mononuclear cells (PBMC) from whole blood or buffy coat preparations by density gradient centrifugation, followed by adhesion-mediated purification of monocytes from lymphocytes. Alternative methods, described later, include centrifugal counterflow elutriation and magnetic negative immunoselection.

Isolation of human monocytes by density gradient centrifugation and adherence

Materials and reagents

Tissue culture plasticware (Falcon)
RPMI 1640, L-glutamine, penicillin/streptomycin (Life Technologies, Paisley, UK)
Ficoll–Hypaque (Pharmacia, Uppsala, Sweden)
0.2 μm Serum Acrodisc (Gelman #4525, Fisher Scientific, Loughborough, UK)
X-Vivo 10 (Bio-Whittaker, Walkersville, MD, USA)
Human blood, buffy coat fraction, screened for HIV and hepatitis B viruses (National Blood Service, Bristol, UK)
E-Toxate endotoxin detection kit (Sigma, Poole, UK)

All other reagents are from Sigma unless stated otherwise.

Solutions required (all sterile)

PBS/GE: Ca^{2+}-, Mg^{2+}-free phosphate buffered saline (PBS) + 10 mM glucose + 2.5 mM EDTA.

Adhesion medium (AM): RPMI 1640 + 7.5% autologous human serum (see below), 2 mM glutamine, 50 U/ml penicillin, 50 µg/ml streptomycin.

Mϕ culture medium (MCM): X-Vivo 10 + 1% autologous human serum (see below), 2 mM glutamine, 50 U/ml penicillin, 50 µg/ml streptomycin.

Isolation of PBMC

1 Decant the buffy coat ($\approx$ 50 ml) into two 50 ml centrifuge tubes. Dilute 1:1 with ice-cold PBS/GE. Mix by gentle inversion.

2 Place 15 ml Ficoll-Hypaque at room temperature (RT) into four 50 ml centrifuge tubes. *Gently* overlay each with 25 ml diluted buffy coat. Centrifuge 20 min at 900g, RT (no brake).

3 Remove all except $\approx$ 5 ml of the supernatant into fresh tubes. This supernatant is used as autologous human 'serum'. During the PBMC washes in steps 5–7, heat inactivate at 56°C for 30 min, chill under running tap water, and centrifuge 5 min at 3200g, 4°C. Decant supernatant from the large white fibrin pellet, and filter through Serum Acrodisc filters. Several filters may be required. As this material has not been clotted, it is not actually serum, but rather fibrin-depleted plasma, which lacks the products released by platelets during clotting. However, we use this as the 'serum' in AM and MCM.

4 Remove the last 5 ml of supernatant and discard. Carefully aspirate the PBMC at the interface with a 5 ml pipette placed just above the interface. Place the cells into fresh tubes, two interfaces per 50 ml tube. Dissociate cells by vigorously pipetting up and down with a 5 ml pipette. Fill tubes with ice cold PBS/GE and spin 7 min at 500g, 4°C (half brake).

5 Aspirate supernatant carefully, as the top part of the pellet may be quite loose. Resuspend in $\approx$ 5 ml ice-cold PBS/GE and pipette up and down vigorously. Then fill with PBS/GE and spin 7 min at 250g, 4°C, with full brake.

6 Repeat the wash in step 6 three to four more times, or until supernatant is clear. Cloudiness indicates the presence of platelets. Since platelets secrete many substances which may activate Mϕ, it is important to minimise platelet contamination.

7 Resuspend thoroughly in cold AM and count. Expect $3-8 \times 10^8$ PBMC per buffy coat, or $5-15 \times 10^5$ per ml whole blood.

Purification of monocytes by adherence

1 Plate PBMC in 75 cm² flasks ($\approx 2 \times 10^8$ cells/flask) or tissue culture treated Petri dishes ($\approx 2 \times 10^8$ cells/14 cm dish). Incubate 45–90 min at 37 °C in a humidified incubator (95% O_2/5% CO_2).
2 Remove AM and non-adherent cells. Wash residual non-adherent lymphocytes from the adherent cell layer with RPMI (NB 37 °C), as follows: add 10–15 ml RPMI, swirl the dish or flask gently, pour off RPMI, repeat 5–8 times. Check by phase contrast microscopy after every 2–3 rinses to ensure that the adherent cells are not beginning to detach. If desired, the AM and non-adherent cells (a mixture of lymphocytes and monocytes) can be replated in a fresh flask and subjected to a second round of adherence to increase the total monocyte yield.
3 Add 15 ml AM per 75 cm² flask. Leave in incubator overnight.
4 To replate monocytes on day 1, place flasks on ice for 15–30 min. The cells will become less adherent, and can be removed by a combination of striking the flask firmly and repeatedly pipetting the medium onto the flask surface with a 5 ml pipette. Pipette cell suspension into a centrifuge tube, rinse the flask with PBS, and collect cells by centrifugation for 7 min at 250g at RT. Count the cells with a haemocytometer. Expect $3-8 \times 10^7$ monocytes per buffy coat. Replate in an appropriate vessel in MCM (*not* AM), and place in a 37 °C incubator.

Alternative methods of monocyte isolation

The adhesion method described above represents the simplest and most commonly used method for monocyte isolation. However, this method is not always the most suitable. The adhesion process itself, mediated by a combination of type 3 complement receptor and the scavenger receptor (Fraser, Hughes & Gordon 1993) and probably others, may lead to Mϕ activation incompatible with the experiment. In addition, while the expected purity of adherence-selected cultures is quite high (at least 90–95%), higher levels of purity can be attained with other methods.

Monocytes can be purified by their characteristic size and density using counterflow centrifugal elutriation. This method generates very pure monocyte populations (up to $\approx 96-98\%$). Moreover, the monocytes are less

activated than those isolated by adherence, since adherence itself may activate the cells. However, elutriation requires expensive and specialised centrifugation equipment, so is unsuitable for many users. A technique for
elutriation of cells collected by leukapheresis of human donors has also been
described, allowing the collection of very large numbers of cells ($>10^9$) from
a single donor (Faradji *et al.*, 1994).

Another alternative to isolation by adherence is negative immunomagnetic selection of lymphocytes from the PBMC population (Flo *et al.*,
1991). Like elutriation, this technique yields very high purity monocytes, at
a low state of activation. Freshly isolated PBMC are treated with magnetic
beads conjugated with antibodies to remove unwanted non-monocytic cells,
for example anti-CD2, -CD19, and -CD56, for removal of T cells, B cells,
and natural killer cells, respectively (Flo *et al.*, 1991). The main drawback to
this technique is the cost of antibodies and magnetic beads. There is little
advantage to positive immunoselection with a monocyte-specific antibody,
since positive immunoselection is likely to activate the monocytes.

Maintaining human macrophages in culture

Monocyte-derived Mϕ can be maintained in culture for at least two weeks.
Typically no change of medium is necessary for the first week, after which
the medium is changed approximately every three days. The cells
differentiate into Mϕ over the first few days in culture, but exhibit an array
of different cell morphologies. Many are non-adherent, though larger and
less round than lymphocytes, and often with characteristic processes and/or
membrane ruffles. Some are rounded and adhere weakly; others are larger
and very strongly adherent and flattened, either maintaining a round shape
(a 'fried egg' morphology), or a more elongated, bipolar or stellate shape.
After 7–10 days in culture, some cells fuse into multinucleated cells. This
morphology is greatly enhanced in the presence of IL-4 or IL-13, resulting
in syncytium-like multinucleated giant cells. Figure 6.2 shows examples of
primary human monocytes/Mϕ after four or 11 days in culture.

Many variations of these culture conditions are possible. If only a few large
vessels are needed, monocytes can be purified from the lymphocytes in the
vessels needed, and placed directly in MCM after washing off the non-
adherent lymphocytes, eliminating the need for replating on the next day.
However, this is not convenient if cells are needed in many smaller vessels,
since washing would be overly tedious.

Monocytes have also been cultured in AM long-term, instead of switching

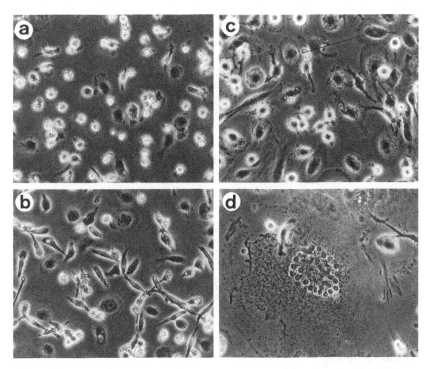

Fig. 6.2. Effect of time in culture and cytokine treatment on human monocyte/macrophage morphology. Human monocytes were isolated by Ficoll gradient and adherence to tissue culture plastic. Cells were cultured for five days (a,b) or 11 days (c,d). Monocytes treated with 200 u/ml IFNγ (b) exhibited a more bipolar morphology, with more vacuoles, than untreated cells (a). Mature Mφ at day 11 (c) were larger and more adherent than those at day five (b). In the presence of IL-13, many of the 11-day Mφ have fused into multinucleated giant cells (d). Magnification of all panels is × 400.

to MCM. These cells tend to acquire many refractile lipid droplets, because of the high lipid content of human serum. Alternatively, RPMI 1640/10% fetal calf serum (FCS) (R10) can be used, which decreases the accumulation of lipid droplets. However, FCS may inhibit differentiation of monocytes. Some morphologies, e.g. multinucleated giant cells, will not be seen in the absence of human serum. Finally, the cells can be cultured completely serum-free in X-Vivo. Again, giant cells are unlikely to appear. In addition, monocytes cultured in serum-free conditions have an increased tendency to die over a period of one to two weeks, probably as a result of apoptosis. If compatible with the particular experimental protocol, this increased cell death can be reduced by addition of Mφ colony-stimulating factor CSF-1 and/or IL-4.

Since the phenotype of Mφ can vary depending on their substrate, the choice of culture vessel can be significant. We routinely use standard tissue culture treated plastic. Other choices include bacterial plastic, which allows for more rapid differentiation; gelatin-coated plastic, which allows easier removal of Mφ from the dish; and glass, to which the Mφ adhere very tightly (Montaner, Collin & Herbein, 1996). Finally, Mφ can be cultured under non-adherent conditions in Teflon bags or bottles (Nalgene, Rochester, USA). Mφ cultured under these conditions are non-adherent or only loosely adherent. Note, however, that some cell functions may be altered.

Because of their strong adhesion, much of which is divalent cation-independent, Mφ can present problems when it is necessary to remove them from their substratum. We have found that placing flasks on ice for 15–30 min, followed by vigorous pipetting with a 5 ml pipette, is often effective. Alternatively, supplement the medium with 4 mg/ml lidocaine-HCl, ±5 mM EDTA. After 10–20 min at 37°C, most cells can be easily removed from the dish.

Troubleshooting

The most commonly cited cause for poor viability and/or yield of human monocyte preparations is donor to donor variability. As a result, often the best strategy is simply to repeat the procedure. If results are consistently unsatisfactory, any of the following actions may improve results: change the substratum (e.g. to gelatin-coated plastic); change the serum concentration of AM or MCM; perform additional washes of PBMC to remove platelets; change the adherence time or washing method of monocyte layers. Finally, check that all media components are endotoxin free, using the E-toxate endotoxin detection kit.

Isolation of other macrophage cell types

The use of monocyte-derived human Mφ has dominated the literature, primarily because of the relative ease of obtaining them. However, differentiated Mφ have been isolated from several other human tissues. In general, these procedures involve enzymatic digestion of a freshly isolated tissue sample, followed by purification of Mφ from other cell types either by adherence or magnetic immunoselection, similar to that described above. Discussion of specific methods is beyond the scope of this chapter.

Applications

Immunocytochemistry

Fluorescence-activated cell sorting (FACS) analysis of monocytes and Mϕ is a powerful tool for the study of antigen expression in response to differentiation, cytokine and chemokine treatment, and exposure to endocytic and phagocytic stimuli. FACS analysis is also a simple method for assessing purity of monocyte cultures, for example by staining with a monocytic marker, such as CD14 or CD68, and a lymphocyte marker not expressed on monocytic cells, such as CD3.

Monocytes and Mϕ purified by the methods described in this chapter can generally be analysed by FACS without gating. In addition, the monocyte portion of PBMC can be analysed without purification, by setting gates to exclude the signal from lymphocytes. When forward scatter and side scatter are measured for a PBMC pool, the monocytes (10–20% of the total events) appear larger and more granular than the lymphocytes (80–90% of total events).

Solutions required

Lidocaine/EDTA (10×): 150 mM lidocaine HCl (Sigma), 50 mM EDTA in PBS.

Paraformaldehyde: 4% paraformaldehyde in PBS. Add 4 g paraformaldehyde and 1 drop 5 M NaOH to 100 ml PBS. Stir with gentle heating until dissolved. Check pH, and filter before use.

PBS/Triton X-100 (PBS-T): 0.2% Triton X-100 in PBS

Blocking buffer (BB): 10% goat serum (or other to match species of secondary antibody) in PBS-T.

Primary antibody (various). Dilute in BB.

Secondary antibody: fluorescein-labelled. Dilute in BB.

1 Add one-tenth volume 10× lidocaine/EDTA and incubate at 37 °C for 15–30 min. Lift off the now loosely adherent cells by pipetting vigorously with a 5 ml pipette. If necessary, a cell scraper may be used, but more debris may be present on FACS analysis. Collect cells by centrifugation (all centrifugations in this protocol are 250g, 5 min).

2 Resuspend the cells in paraformaldehyde (10^5–10^6 cells/ml) and fix for 20 min (all steps are performed at RT). Centrifuge.

3 Permeabilise the cells by resuspending in BB at 3–10×10^5 cells/ml. Incubate 20 min. If permeabilisation is not desired, resuspend in PBS and

use PBS instead of PBS-T in all other solutions. Pipette 100 μl aliquots into the wells of a round-bottom 96-well plate (Falcon). Centrifuge.

4 Remove supernatant with one 'flick' of the plate. Resuspend cells in 100 μl of primary antibody. Incubate 30 min. Centrifuge.

5 Wash cells twice by resuspending in 200 μl PBS-T followed by centrifugation.

6 Resuspend cells in 100 μl secondary antibody. Incubate 30 min, then wash twice as in step 5.

7 Resuspend cells in 150 μl PBS-T. The cells are now ready for FACS analysis. If analysis cannot be performed immediately, store cells at 4 °C in the dark.

Figure 6.3 shows expression of monocytic markers in human monocyte/Mφ after 1, 4, and 11 days of culture. The pan-monocytic cell marker CD68 is highly expressed throughout, while the T cell-specific marker CD3 is negative throughout. Expression of HLA-DR increases with time in culture, while CD16 expression decreases slightly. Macrophage mannose receptor (MMR) is barely detectable in monocytes, but becomes strongly expressed as the monocytes differentiate into Mφ. Expression of the commonly used monocytic marker CD14 (not shown) is high in monocytes, but variable in Mφ depending on the culture conditions.

Since Mφ express Fc receptors that bind immunoglobulins, it is critical that antibody treatments be performed in the presence of serum from the same species as the labelled (secondary) antibody. Alternatively, potential problems with Fc receptor binding can be avoided by using F(ab')$_2$ fragments instead of whole antibodies. Since fixation reduces Fc receptor activity, non-specific Fc receptor binding is more problematic when staining live cells. Controls for non-specific staining are either no primary antibody or an isotype-matched irrelevant antibody for monoclonals, and pre-immune or non-immune serum for polyclonals. These controls are critical, since Mφ often display significant autofluorescence.

Multinucleated giant cell fusion assay

Materials

8-well glass chamber slides (Nunc, Naperville, IL, USA)
MCM medium (see above)
4% paraformaldehyde solution (see above)
IL-4 or IL-13 (R&D Systems, Abingdon, UK)
Haematoxylin and eosin (Sigma)

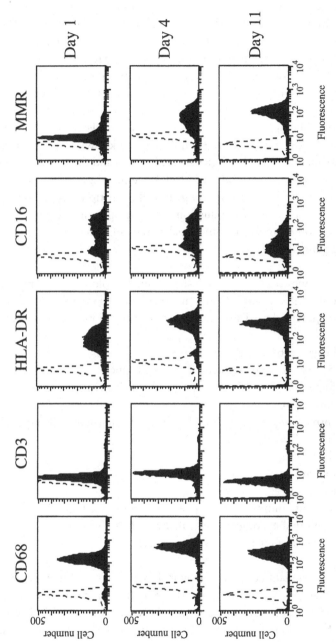

Fig. 6.3. FACS analysis of human macrophage markers. Human monocytes were isolated by Ficoll gradient and adherence to tissue culture plastic. Cells were cultured for one, four, or 11 days. Cells were collected using lidocaine/EDTA, then stained with monoclonal antibodies against CD68, CD3, HLA-DR, CD16, and mannose receptor (a kind gift of Dr D. C. Rijken, Leiden, The Netherlands). In control experiments (dashed lines) primary antibody was omitted.

In the presence of autologous human serum and IL-4 or IL-13, Mϕ undergo a process of clumping, followed by cell fusion, resulting in the formation of multinucleated giant cells. Giant cells are large and extremely flat, and may contain 100 or more nuclei. While their functional relevance is unclear, cells with very similar morphology exist *in vivo* under certain pathologic conditions. Mϕ-derived giant cells have been implicated in the pathology of the fatal heart disease, giant cell myocarditis. In addition, giant cells form at sites of tissue injury and around arterial grafts composed of synthetic materials. The presence of these giant cells is correlated with chronic inflammation. At least two cell surface molecules are known to be involved in giant cell formation, CD98 (Ohgimoto *et al.*, 1995) and the Mϕ mannose receptor (McNally, DeFife & Anderson, 1996). Despite these advances, very little is known about the mechanism of giant cell formation. The following protocol describes a method for induction and assay of giant cell formation. This assay is a good example of a functional activity of Mϕ that can only be studied using primary cultures.

1 Remove day 1 monocytes from their substratum by cold shock and replate onto eight-well glass chamber slides in MCM plus 10 ng/ml IL-4 or IL-13. Add any potential mediators (e.g. antibodies, cytokines, inhibitors) to duplicate wells.
2 Every 2–3 days, re-treat with fresh cytokine.
3 By day 4–5, large rounded clusters of cells will appear. By day 6–8, giant cells will form. Fix the cells with 4% paraformaldehyde and stain with haematoxylin/eosin.
4 Carefully inspect the slides under a bright field microscope. Score each well using the following scale:

Score	Phenotype
0	no giant cells
1	few giant cells
2	moderate numbers of giant cells, few with >20 nuclei
3	large numbers of giant cells, many with >20 nuclei
4	most of the nuclei are inside giant cells

An example of a giant cell fusion assay is shown in Fig. 6.4 (see also Fig. 6.2d for an example of a giant cell). IL-13 induces formation of large numbers of giant cells in these 6-day Mϕ cultures. IL-4 is equally effective (not shown). Formation of giant cells is dependent on the presence of human serum, with FCS having little effect.

It is also possible to quantitate giant cell formation by generation of a

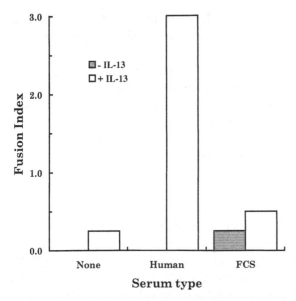

Fig. 6.4. Effect of IL-13 and serum on giant cell formation. Human monocytes were isolated by Ficoll gradient and adherence to tissue culture plastic. Cells were transferred onto glass chamber slides and cultured for 6 days in X-Vivo with no serum, 1% autologous human serum, or 1% FCS, in the presence or absence of 10 ng/ml IL-13. Slides were stained with haematoxylin/eosin and a fusion score assigned as described in the text. Data are the mean of duplicate wells.

fusion index. High-power fields are selected at random, and the percentage of nuclei inside giant cells calculated. However, we have found that giant cells are often non-randomly distributed in the wells, with more giant cells occurring at the edges. Therefore, the fields chosen for generation of a fusion index may not accurately represent the distribution of cells in the well.

Macrophage gene expression

Primary cultures of monocyte/Mϕ are our most valuable tool in the study of how Mϕ alter their gene expression patterns to perform a wide variety of different functions. To understand the mechanism of Mϕ differentiation and activation, we must first understand the proteins switched on or off by these processes. While some of these studies can be done using immunocytochem- ical techniques, perhaps the most powerful technique is the study of gene expression at the mRNA level. The most rapid and sensitive method for studying expression of known genes is reverse transcriptase-polymerase chain reaction (RT-PCR). Differential display (DD)-PCR is an adaptation of this technique for the identification of differentially expressed unknown genes.

The techniques for the study of tissue-specific and 'activation-dependent' gene expression are relatively constant across all tissues. Therefore, rather than repeating standard methods, we will concentrate on the specific application of these methods to primary human Mϕ.

Isolation of RNA from primary human macrophages

Materials

RNAzol (Biogenesis, Poole, UK)
QuickPrep micro mRNA Purfication Kit (Pharmacia)

Purification of high-quality RNA is a critical step for successful RT-PCR and DD-PCR. It is important to decide whether the entire population of a culture, or just the adherent portion, is to be collected. While the non-adherent portion is somewhat more likely to contain non-Mϕ, in most cases the non-adherent pool probably represents a valid alternate Mϕ phenotype, and not merely a contaminant. Therefore, discarding this pool risks losing potentially novel Mϕ-specific sequences.

We have had good success preparing total RNA using RNAzol, and preparing messenger RNA with the QuickPrep micro mRNA purification kit. In either case, the medium is removed and non-adherent cells collected by centrifugation at 250g for 5 min. Meanwhile, RNAzol reagent or the mRNA extraction buffer is added directly to the culture flask. The flask is scraped with a cell scraper to lyse all adherent cells. All of the supernatant is carefully removed from the non-adherent cells, and the pellet is dislodged. The same extraction buffer is transferred from the flask to the centrifuge tube, thus creating one lysate from both adherent and non-adherent cells. The remainder of the RNA isolation is performed according to the manufacturers' protocols.

Differential display PCR

We have performed DD-PCR, using random 10mers paired with anchored oligo(dT) as primers, according to the originally published method (Liang & Pardee, 1992). DD-PCR often produces false positives, so it is important to set stringent criteria for choosing differentially expressed products to pursue. While in its simplest form only two mRNA pools are compared, the use of more complex designs allows a more restrictive selection protocol. For example, a dual requirement for both Mϕ-specific expression and regulation by one or more treatments (e.g. cytokine treatment, time in culture, or phagocytic stimulus) is preferable. If the experimental question only allows a single selection parameter, it is best to perform DD-PCR reactions in duplicate from independent mRNA pools. Only products with consistent differential expression should be selected. Clone the excised, re-amplified products of interest into an appropriate sequencing vector and perform DNA sequence analysis. Compare the sequences obtained to the Genbank and EST databases. Re-screen individual clones of interest for differential expression by Northern blot or screen multiple clones by reverse Northern blot (Poirier et al., 1997).

Isolation and culture of primary murine macrophages

Introduction

The mouse has provided a convenient and reliable source of cells for studying monocytic development and function. Whilst there are differences between murine and human monocyte/Mϕ, there are many similarities. For example, the discovery of nitric oxide (NO) as a potent cytocidal Mϕ secretory product was originally made in the mouse, but inducible nitric oxide synthase has recently been confirmed in alveolar Mϕ from patients with tuberculosis (Nicholson et al., 1996).

Recently the development of transgenic and gene knockout mice has provided new tools for the study of Mϕ biology and the role of Mϕ in host defence. In addition, rapid progress is now being made in elucidating the molecular biology underlying many aspects of Mϕ function. Therefore it seems likely that the mouse will continue to be a valuable resource for the investigation of the diverse functions of the cells of the Mϕ lineage.

In contrast to man, the mouse enables not only PBMC but also resident Mϕ to be readily isolated. The injection of specific inflammatory agents into the peritoneal cavity elicits a Mϕ-rich cell population, which can be easily harvested with high yield. By varying the stimulus, the investigator can recruit cells with varying phenotypes and activation states. This section will describe the isolation and culture of a wide range of primary murine Mϕ.

Isolation of primary murine macrophages

Peritoneal macrophages

The peritoneal cavity is a convenient source of primary Mϕ, which can subsequently be purified through an adhesion step. Peritoneal Mϕ can be divided phenotypically into resident, elicited and activated cells according to stimulus. These cells are harvested following peritoneal lavage, as outlined below.

Resident peritoneal macrophages (RPM)

Sacrifice a mouse by CO_2 inhalation and pin to a dissection board with the ventral surface exposed. Wet the skin with 70% ethanol and, taking care not to cut into the body wall, make a small cut in the skin over the abdomen using fine scissors. Retract the skin from the incision to reveal the shiny surface of the body wall. This area should remain sterile during the entire procedure. Then, using a 21 gauge needle, slowly inject approximately 10 ml of sterile PBS into the peritoneal cavity. The needle should be inserted bevel uppermost into the caudal half of the abdomen, taking care not to penetrate internal viscera. As the needle is removed there may be some loss of saline, however omental fat usually blocks further leakage. Remove the pins from the forelegs of the mouse and gently shake the entire body for ≈ 10 s. Do not shake too vigorously, otherwise haemorrhage will result in red cell contamination of the cell harvest. Slowly withdraw the saline containing suspended cells from the peritoneal cavity, using a 19 gauge needle attached to a 10 ml syringe. The needle should be inserted bevel down into the cranial half of the cavity to avoid fat blocking the needle during aspiration. Store the cell suspension, which is typically 25–40% Mϕ, on ice until required. Mϕ may be purified by adhesion, as described below for Biogel-elicited cells.

Biogel polyacrylamide bead elicited cells (BgPM)

Materials

Biogel P-100 polyacrylamide beads (fine; hydrated size 45–90 µm; Bio-Rad, Richmond, CA, USA)
RPMI 1640 or other medium (see below), Life Technologies

In this laboratory Biogel beads have been used successfully to elicit a high yield of peritoneal Mϕ (10^7 cells per mouse). These beads cannot be digested or phagocytosed by Mϕ. Since the resulting cells are free of debris, they are especially suitable for studies of endocytosis.

Wash Biogel beads several times in PBS, and autoclave before use. Inject mice intraperitoneally (i.p.) with 1 ml of a 2% v/v suspension of beads. Harvest the cells on day 4 or 5, as outlined for RPM above. Both cells and beads will be collected. Plate the cells in RPMI 1640 plus 10% FCS at $3\times$

10^5 per well in a 24-well tissue culture plate and incubate for 60 min at 37 °C. Remove beads and non-adherent cells (mostly neutrophils) by washing the wells five times with PBS. The adherent cells consist of greater than 90% Mϕ, and viability is usually greater than 97% by phase contrast microscopy and trypan blue exclusion.

Biogel-elicited Mϕ (BgPM) have a number of distinct phenotypic features. For example, after overnight incubation in serum-containing medium on tissue culture plastic (TCP), over 50% of BgPM become non-adherent and the remainder are rounded up. In contrast, thioglycollate-elicited Mϕ (TPM), which are discussed below, remain flattened and firmly adherent to the plastic (M. Stein, unpublished observations).

Thioglycollate broth elicited cells (TPM)

Materials

Brewer's complete thioglycollate broth (Difco Laboratories, Surrey, UK)

Harvest thioglycollate Mϕ (TPM) from the peritoneal cavity 4 to 5 days after i.p. injection of 1 ml of Brewer's complete thioglycollate broth. Note that thioglycollate broth often contains low levels of lipopolysaccharide (LPS) (0.5 ng/ml), which may affect Mϕ responsiveness in subsequent assays. Proteose-peptone can be used as a similar stimulant but recruits fewer Mϕ and they contain fewer phagocytic vacuoles.

BCG recruited cells

Materials

Mycobacterium bovis Bacille Calmette-Guérin (BCG), Pasteur strain (kindly provided by Dr Genevieve Milon, Pasteur Institute, Paris, France)

Intraperitoneal injection of 10^7 colony forming units (CFU) of BCG organisms causes recruitment of immunologically activated Mϕ to the peritoneal cavity (Ezekowitz et al., 1981). BCG stocks are stored at -70 °C, resuspended in PBS and sonicated just before use. Harvest peritoneal Mϕ by lavage as described for RPM above, 4 to 6 days following injection. Percoll gradients can be used at this stage to enrich the population for Mϕ (Pertoft & Laurent, 1977).

BCG Mϕ become activated *in vivo* under the influence of T cell cytokines, such as IFNγ, and express high levels of cell surface MHC Class II (Fig. 6.1). These IFNγ primed cells represent a highly responsive population that can be used, for example, to investigate the activated Mϕ response to bacterial cell products such as LPS and lipoteichoic acid. Changes in expression of cell surface and intracellular antigens can provide useful information regarding activation status or endocytic activity of Mϕ (Fraser *et al.*, 1994). FACS analysis allows us both to identify what types of cells are present in a mixed population, such as BCG-recruited cells, and also to analyse the activation status of the Mϕ in the population. In the example shown in Fig. 6.5, approximately 30% of the peritoneal cells are Mϕ (macrosialin positive). The cells are activated, as shown by high levels of MHC Class II and 7/4 expression (confirmed by two-colour staining, not shown). They do not express sialoadhesin, a marker of stromal Mϕ.

Peripheral blood mononuclear cells (PBMC)

Materials

Nycoprep 1.077 Animal (Nycomed Pharma AS, Norway)
Tris buffered ammonium chloride (TBAC) lysis buffer:
  A. 0.15 M ammonium chloride
  B. 0.17 M Tris base
  Mix solutions in a ratio of 9:1 A:B and adjust pH to 7.2

To harvest PBMC, collect cardiac blood into a heparinised syringe using a 25-gauge needle. Add an equal volume of 0.9% saline and layer the diluted blood onto a Nycoprep cushion. Centrifuge at 900g (no brake) for 15 min. Harvest the mononuclear cell-rich fraction from the interface between the plasma and the Nycoprep cushion. Lyse erythrocytes by resuspending the cells in TBAC lysis buffer. Collect the remaining cells by centrifugation and resuspend in RPMI 1640 plus 10% FCS prior to use.

Tissue macrophages

Resident tissue Mϕ can be isolated from a range of different murine organs using the enzymatic methods described below. In addition, activated Mϕ can be isolated from infected or inflamed tissues.

Spleen macrophages

Materials

Collagenase D, DNAse I (Boehringer Mannheim, Lewes, East Sussex, UK)
Cell strainers, 70 μm (Falcon)

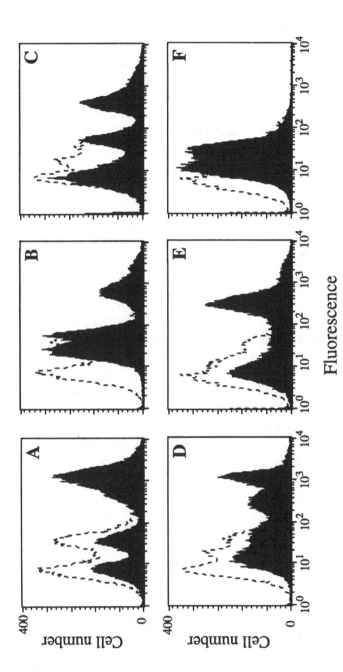

Fluorescence

Fig. 6.5. FACS analysis of BCG-recruited peritoneal cells. Peritoneal cells were recruited following i.p. inoculation of Sv/129 mice with 10⁷ cfu of BCG. Cells were harvested seven days later and, prior to purification, were stained with primary antibodies against the following antigens: (A) MHC Class II; (B) class A Mφ scavenger receptor; (C) macrosialin, a Mφ specific marker; (D) 7/4 antigen, a marker of neutrophils and activated Mφ; (E) type 3 complement receptor, expressed on Mφ and neutrophils; (F) sialoadhesin, a product limited to stromal Mφ. Cells in all panels except (E) and (F) were permeabilised. Dashed lines represent staining of isotype-matched control antibodies.

Remove spleens and store in PBS on ice. Digest the spleens in 0.05% collagenase and 0.002% DNase in RPMI 1640 without serum at 37°C for 30 min. Disrupt organ fragments by vigorous pipetting and filter the suspension through a cell strainer. In order to maintain Mϕ cell integrity, it is important that sufficient enzymatic digestion has taken place before mechanical disruption is used. Purify the Mϕ by adhesion as described above for BgPM. Mϕ in the spleen represent a heterogeneous population and different approaches may be used to isolate Mϕ sub-populations (Dijkstra *et al.*, 1985).

Resident bone marrow macrophages (RBMM)

Materials

Collagenase D, DNAse I (Boehringer Mannheim)
RPMI 1640 (Life Technologies)
Ficoll–Hypaque (Pharmacia)
25 gauge needles

Sacrifice mice and sterilise the abdomen and hind legs with 70% ethanol. Dissect the skin away from the abdomen and hind legs. Using fine scissors, remove the muscles attaching the hind limb to the pelvis and those attaching the femur to the tibia. Cut through the tibia just below the knee joint using strong scissors, and free the femur from the mouse by cutting through the pelvic bone close to the hip joint. Store the femurs in RPMI 1640 on ice prior to use. Place the bones in 70% ethanol for 1 min to maintain sterility, before washing twice with PBS.

Hold each femur firmly with forceps, and, with a single motion, cut off the expanded ends (epiphyses) using strong scissors. Flush out the bone marrow by forcing 5 ml of RPMI 1640 containing 0.05% collagenase and 0.001% DNase through the central cavity using a syringe attached to a 25 gauge needle. Suspend the marrow from two femurs in 10 ml of the same enzyme solution and digest at 37°C for 1 h with shaking. Stop digestion by adding FCS to a final concentration of 1% v/v. The marrow plug fragments should have dispersed and yielded a homogenous cell suspension.

Enrich purified cell clusters by gravity sedimentation in RPMI containing 30% FCS or by use of a Ficoll–Hypaque cushion. Wash the clusters twice in RPMI by centrifugation at 100g for 10 min, resuspend in R10 and add to tissue culture plates. After 2 h incubation at 37°C, remove non-adherent cells by washing with PBS. The cells remaining have a characteristic morphology and represent resident bone marrow Mϕ (RBMM), contaminated with some monocytes and neutrophils.

Bone marrow derived macrophages (BMDM)

Materials

RPMI 1640 (Life Technologies)
Cell strainers, 70 μm (Falcon)
TBAC lysis buffer (see above)
5 mM EDTA in PBS
recombinant CSF-1 (R&D Systems)

BMDM are not mature Mϕ, but rather non-adherent Mϕ precursors. Flush femurs with RPMI (no enzymatic digestion is required). Pass the resulting cell suspension through a cell strainer and centrifuge at 300g for 5 min. Lyse erythrocytes by resuspending in TBAC lysis buffer for 5 min at RT and wash the remaining cells three times in RPMI 1640.

Resuspend 5×10^5 cells/ml in RPMI 1640 plus 10% FCS and recombinant CSF-1 and culture on bacterial plastic (BP) plates. Alternatively, L cell conditioned media can be used as a source of CSF-1 (Hume & Gordon, 1983). After 3 days of incubation, wash adherent cells and add fresh medium. After 7 days, detach the cells with 5 mM EDTA in PBS. These cells are mature, proliferating Mϕ, capable of mediating phagocytosis of a number of different particles and suitable for use in a wide range of assays.

Fetal liver macrophages

Materials

Collagenase D, DNAse I (Boehringer Mannheim)

The most abundant source of Mϕ in the developing mouse fetus is the liver. Fetal liver Mϕ and attached erythroid cell clusters can be isolated from day 14 fetal liver by collagenase/DNase digestion followed by adhesion to plastic (Morris, Crocker & Gordon, 1988).

Culture of macrophages

Culture Mϕ in a humidified incubator at 37 °C in an atmosphere containing 5% carbon dioxide. The conditions selected for culture of primary murine Mϕ, such as media and tissue culture flask, will depend on which assays are planned. Mature murine Mϕ will not divide *in vitro* in the absence of growth factors such as CSF-1; however, under the appropriate conditions, they may remain viable for up to five days.

Troubleshooting

While the isolation and culture of human monocyte/Mϕ can be difficult, isolation of murine peritoneal macrophages, whether resident or elicited as described above, rarely presents a problem. The elicited methods circumvent the main problem associated with resident peritoneal macrophages, i.e. a low overall yield. Depending on the substrate used, the yield and purity of elicited peritoneal macrophages may vary somewhat, but the proper balance of yield versus purity can be adjusted by varying the number and force of washes after cell adhesion. The degree of activation of the macrophages can be tailored to the specific application by using different stimulations: BCG-recruited cells are more active than thioglycollate-recruited cells, which in turn are more active than the mostly quiescent Biogel-elicited cells.

Media and sera

Mϕ can be cultured in a number of different types of media such as Minimum Essential Medium (MEM), RPMI 1640 and Dulbecco's modified Eagles medium (DMEM) (Life Technologies). Supplement the media with 5–10% FCS, 2 mM glutamine, 50 U/ml penicillin and 50 µg/ml streptomycin (Life Technologies). RPMI 1640 is also usually supplemented with 10 mM Hepes (pH 7.3) for additional buffering. Heat-inactivate FCS at 56 °C for 30 min and filter sterilise (0.22 µm pore size) prior to use.

Proprietary serum-free media, such as Optimem (Life Technologies), can be useful when the investigator wishes to exclude the effect of serum components in a particular assay. However, Mϕ adhere tightly to plastic substrata under serum-free conditions by a poorly understood cation-independent mechanism (Fig. 6.6). This strong adhesion creates difficulties in harvesting the cells from their substrata prior to their use in assays.

Substratum

Murine Mϕ adhere firmly to their substrata when cultured on TCP (tissue culture plastic) or BP (bacteriologic plastic) in the presence of serum. BP-cultured Mϕ detach after incubation in PBS plus 5 mM EDTA for 30 min, while TCP-cultured Mϕ require EDTA/lidocaine treatment for detachment (see above). Trypsin is ineffective at removing Mϕ from this surface. Wash cells thoroughly with fresh media after EDTA/lidocaine treatment. Teflon coated tissue culture vessels have proved useful for culturing murine Mϕ in suspension.

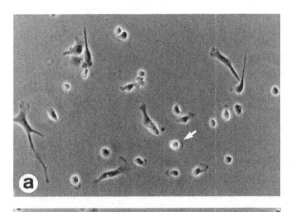

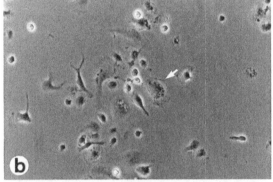

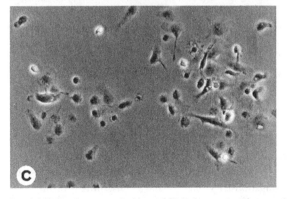

Fig. 6.6. Spreading morphology of Biogel-recruited macrophages (BgPM) in a variety of culture media. Peritoneal cells were recruited following i.p. inoculation of Sv/129 mice with Biogel beads. Cells were harvested five days later and allowed to adhere for 6 h in (a) R10; (b) RPMI without serum; (c) Optimem, a proprietary serum-free medium. Non-adherent cells were removed by repeated washing with PBS and the remaining cells fixed with paraformaldehyde. Many BgPM failed to spread fully after 6 h in the presence of serum (a), however the cells frequently extended a ruffling edge (arrow, a). In contrast, BgPM spread rapidly in the absence of serum (arrow, b) with many cells containing clearly visible vacuoles. Magnification × 400.

Summary

As an effector cell involved in both innate and acquired immune responses, the monocyte/Mφ performs an unusually diverse set of functions. Its unique phenotypic plasticity allows the monocyte to differentiate into many different forms in its various target tissues throughout the body. When an appropriate signal is given, the resident Mφ is ready to respond in a wide variety of ways: phagocytosis, endocytosis, secretion of soluble signalling factors, and presentation of antigen, to name a few. It is fortunate for researchers in the field that in both the murine and human systems, an abundant and readily purified source of cells is available. The continued use of these primary culture systems, particularly in conjunction with DD-PCR and similar gene expression techniques, should allow continued acceleration in the progress of Mφ biology.

Acknowledgements

Work in the laboratory of S.G. is supported in part by the Medical Research Council. J.A.M. is supported by the British Heart Foundation. R.H. is supported by the Wellcome Trust.

References

Dijkstra, C. D., Van Vilet, E., Dopp, E. A., Van der Lely, A. A. & Kraal, G. (1985). Marginal zone macrophages identified by a monoclonal antibody: characterisation of immuno- and enzyme-histochemical properties and functional capacities. *Immunology*, **55**, 23–30.

Ezekowitz, R. A. B., Austyn, J., Stahl, P. & Gordon, S. (1981). Surface properties of Bacillus Calmette-Guérin-activated mouse macrophages. Reduced expression of mannose-specific endocytosis, Fc receptors, and antigen F4/80 accompanies induction of Ia. *J. Exp. Med.*, **154**, 60–76.

Faradji, A., Bohbot, A., Schmittgoguel, M., Siffert, J. C., Dumont, S., Wiesel, M. L., Piemont, Y., Eischen, A., Bergerat, J. P., Bartholeyns, J., Poindron, P., Witz, J. P. & Oberling, F. (1994). Large-scale isolation of human blood monocytes by continuous-flow centrifugation leukapheresis and counterflow centrifugation elutriation for adoptive cellular immunotherapy in cancer patients. *J. Immunol. Methods*, **174**, 297–309.

Flo, R. W., Naess, A., Lundjohansen, F., Maehle, B. O., Sjursen, H., Lehmann, V. & Solberg, C. O. (1991). Negative selection of human monocytes using magnetic particles covered by antilymphocyte antibodies. *J. Immunol. Methods*, **137**, 89–94.

Fraser, I., Hughes, D. & Gordon, S. (1993). Divalent cation-independent macrophage adhesion inhibited by monoclonal-antibody to murine scavenger receptor. *Nature*, **364**, 343–5.

Fraser, I. P., Doyle, A. G., Hughes, D. A. & Gordon, S. (1994). Use of surface molecules and receptors for studying macrophages and mononuclear phagocytes. *J. Immunol. Methods*, **174**, 95–102.

Gordon, S. (1996). Overview: the myeloid system. In *Weir's Handbook of Experimental Immunology*, 5th edn, vol. IV, ed. L. A. Herzenberg, D. M. Weir, L. A. Herzenberg & C. Blackwell, pp. 153.1–.9. Cambridge, USA: Blackwell Science.

Gordon, S., Clarke, S., Greaves, D.R. & Doyle, A. (1995). Molecular immunobiology of macrophages: recent progress. *Curr. Opin. Immunol.*, **7**, 24–33.

Hume, D. A. & Gordon, S. (1983). Optimal conditions for proliferation of bone marrow-derived mouse macrophages in culture: the roles of CSF-1, serum, calcium, and adherence. *J. Cell. Physiol.*, **117**, 189–94.

Liang, P. & Pardee, A. B. (1992). Differential display of eukaryotic messenger RNA by means of polymerase chain reaction. *Science*, **257**, 967–71.

McNally, A. K., DeFife, K. M. & Anderson, J. M. (1996). Interleukin-4-induced macrophage fusion is prevented by inhibitors of mannose receptor activity. *Am. J. Pathol.*, **149**, 975–85.

Montaner, L. J., Collin, M. & Herbein, G. (1996). Human monocytes: isolation, cultivation, applications. In *Weir's Handbook of Experimental Immunology*, 5th edn, vol. IV, ed. L. A. Herzenberg, D. M. Weir, L. A. Herzenberg & C. Blackwell, pp. 155.1–11. Cambridge, USA: Blackwell Science.

Morris, L., Crocker, P. R. & Gordon, S. (1988). Murine fetal liver macrophages bind developing erythroblasts by a divalent cation-dependent hemagglutinin. *J. Cell Biol.*, **106**, 649–56.

Nicholson, S., Bonecina-Almeida, M., Silva, J. & Ho, J. L. (1996). Inducible nitric oxide synthase in pulmonary alveolar macrophages from patients with tuberculosis. *J. Exp. Med.*, **183**, 2293–302.

Ohgimoto, S., Tabata, N., Suga, S., Nishio, M., Ohta, H., Tsurudome, M., Komada, H., Kawano, M., Watanabe, N. & Ito, Y. (1995). Molecular characterization of fusion regulatory protein-1 (FRP-1) that induces multinucleated giant cell formation of monocytes and HIV GP160-mediated cell fusion – FRP-1 and 4F2/CD98 are identical molecules. *J. Immunol.*, **155**, 3585–92.

Park, D. R. & Skerrett, S. J. (1996). IL-10 enhances the growth of *Legionella pneumophila* in human mononuclear phagocytes and reverses the protective effect of IFN-gamma – differential responses of blood monocytes and alveolar macrophages. *J. Immunol.*, **157**, 2528–38.

Pertoft, H. & Laurent, T. C. (1977). *Methods of Cell Separation*, ed. N. Catsimpoolas, p. 25. New York: Plenum.

Poirier, G. M.-C., Piyati, J., Wan, J. S. & Erlander, M. G. (1997). Screening differentially expressed cDNA clones obtained by differential display using amplified RNA. *Nucleic Acids Res.*, **25**, 913–4.

van Furth, R. (1992). Production and migration of monocytes and kinetics of macrophages. In *Mononuclear Phagocytes: Biology of Monocytes and Macrophages*, ed. R. v. Furth, pp. 3–12. Dordrecht, The Netherlands: Kluwer Academic Publishers.

7

NK cells and LAK cells

Colin G. Brooks and Hergen Spits

Introduction

Natural killer (NK) cells are a population of immune effector cells that account for about 5% of circulating lymphomyeloid cells in man, rodents, and other vertebrate species that have been examined. They were discovered in the mid 1970s during studies on the cytolytic activity of human blood mononuclear cells and mouse spleen cells against tumour cells *in vitro*, and are still best known for their spontaneous (*ex vivo*) cytolytic activity against appropriate tumour target cells. Surprisingly, after more than 20 years of study, NK cells remain an enigmatic and poorly understood component of the immune system, with many of the key questions regarding these cells, in particular their origin and relationship to other cells of the immune system, their mechanism(s) of recognition, and their physiological function, remaining largely unanswered.

With regard to their function, many possibilities have been considered. Following the route of their discovery, it was widely believed that they provided an innate and vital immunosurveillance mechanism against cancer. The discovery that NK cells could be activated by a variety of mediators to even stronger anti-tumour cell cytocidal activity raised the possibility that NK cells might be used therapeutically for the control or eradication of established tumours. Recently, attention has focused more strongly on the likelihood that NK cells play an important role in innate immune responses to infection. Growing evidence suggests that their ability to produce cytokines such as γIFN and TGFβ may be more important in these situations than their cytolytic activity.

As in other areas of immunology, the key to understanding the physiological function of NK cells will undoubtedly come from an understanding of their recognition mechanisms and capabilities. Stimulated by the realisation that at least under certain experimental conditions, and contrary to previous beliefs, NK cells are MHC restricted, substantial progress has been made in this area in recent years. The most surprising and important finding has been that, in contrast to T cells, MHC restriction in NK cells is mediated by receptors that deliver negative (inhibitory) signals following interaction with target cell class I molecules. This mechanism is thought to endow NK cells with the capacity to distinguish normal healthy tissue cells from 'diseased' cells that are lacking the normal complement of MHC/peptide complexes, either because peptides derived from intracellular pathogens have displaced the normal self peptides, because a pathogen has subverted the class I loading pathway in order to evade detection by T cells, or because an uninfected cell has undergone mutation (potentially oncogenic) that affects the display of normal self peptide/MHC complexes at the surface.

These advances in our understanding of NK recognition have been made possible, at least in part, through advances in culture procedures, particularly in the ability to obtain and grow pure populations of NK cells in long-term culture, and most importantly in the cloning of these cells. In this latter regard, progress has been more rapid in humans than in animals. The derivation of NK cell clones from blood mononuclear cells led directly to the discovery of killer inhibitor receptors in the human (Pantaleo et al., 1988). Human NK cell clones can be readily obtained from other anatomical sites such as the thymus (Hori & Spits, 1991). In addition, clones of putatively immature NK cells lacking CD56 and cytolytic activity against NK-sensitive targets and sharing characteristics with immature T cells can be obtained from human fetal liver (Hori et al., 1992). Interestingly, these clones proliferate not only in response to IL-2 but also to IL-3, IL-4, and IL-7, whereas $CD56^+$ fetal blood-derived NK clones responded to IL-2 and IL-4 but not to IL-3 or IL-7 (for a review of the relationship between T cells and NK cells see Spits, Lanier & Phillips, 1995).

For unknown reasons, and in marked contrast to the situation for both human NK cells and mouse T cells, no means has yet been found to induce NK cells from normal adult mice to grow in long-term culture. Thus, although mouse NK cells show strong initial proliferative responses to IL-2, growth ceases after about 2–3 weeks. The resulting quiescent NK cells are capable of surviving for several more weeks if re-fed regularly with fresh IL-2-containing medium, but during this period they usually transform into giant cells that morphologically and functionally bear little resemblance to the starting population. The factors that control this transformation are unclear, although in at least one situation prostaglandins, perhaps released by contaminating macrophages, may be responsible (Linnemeyer & Pollack,

1993). Therefore, despite intensive investigation in many laboratories using a wide range of growth factors and stimulants, including IL-15, which like IL-2 is a potent inducer of short-term proliferation in NK cells, no means has yet been found to generate long-term lines and clones of NK cells from normal adult mice. However, as discussed below, it is now possible to obtain NK cell clones either from mice having a targeted mutation in the p53 tumour-suppressor gene, or, more generally, from fetal mice. In this chapter we will describe protocols that can be used for both the short-term and long-term culture of human and mouse NK cells. However, before discussing these methods it is necessary to deal with the issue of LAK cells and what exactly an NK cell is.

NK cells and LAK cells, a problem of definition

The term lymphokine activated killer (LAK) cell was coined initially to describe those cells of uncertain lineage that, following exposure of human blood mononuclear cells or mouse spleen cells to high concentrations of IL-2, displayed exceptionally high cytolytic activity against 'NK-resistant' tumour cells. The implication that LAK cells are a distinct population of immune cells has persisted to the present day. However, a considerable body of evidence now makes it quite clear that LAK cells are, on the whole, simply activated NK cells (Herberman et al., 1987). Whether there is actually any change in specificity as a consequence of this activation is controversial, but in at least most cases the ability of LAK cells to kill target cells that are resistant to fresh ex vivo NK cells can be explained by a combination of two things: (a) the enrichment of NK cells in the effector population caused by their rapid proliferation in high-dose IL-2; and (b) the great increase in the cytolytic activity of individual NK cells induced by IL-2- and/or IL2-induced secondary mediators.

The exception to the rule that LAK cells are activated NK cells is that many types of T cell, especially cloned T cells, can reversibly acquire NK activity (Pawelec et al., 1982; Brooks, 1983; Shortman, Wilson & Scollay, 1984), either because they have been initially activated by some stimulus in vivo or in vitro and thereby acquired responsiveness to IL-2, or because these populations of T cells constitutively express appropriate form(s) of IL-2 receptors. Numerically, the most prominent T cell contributors to the LAK phenomenon are conventional CD8 T cells (Ballas & Rasmussen, 1990), but under some circumstances the relatively rare 'non-conventional' populations of T cells may contribute substantially, especially when conventional T cells have been depleted from the starting population. These include $\gamma\delta$ T cells (Patel *et al.*, 1989) and CD4⁻CD8⁻ $\alpha\beta$ T cells (Koyasu, 1994), some of which express NK cell markers such as CD16, CD56, and killer inhibitory receptors (KIR) and are often named NK-T cells. However, the definition of what constitutes an NK-T cell and whether NK-T cells constitute a single discrete subset of T cell or (at least in some cases) are T cells that have simply acquired NK cell markers as a consequence of some immunoregulatory event, is uncertain. The fact that these various populations of T cells express both NK cell and T cell recognition molecules may provide a third explanation of why LAK cells have sometimes been reported to have a different specificity from fresh NK cells.

In this chapter, we shall not be concerned with T cells at all, even when they express NK-like activities. However, the above considerations raise two important problems: the conceptual problem of how to define NK cells and the practical issue of how to obtain pure populations and cultures of these cells. The conventional definition of NK cells as 'large granular lymphocytes that lack T cell and B cell receptor gene rearrangements and display spontaneous non-MHC-restricted cytotoxic activity' is of little practical value for identifying or purifying NK cells. The major problem is that there are no unique markers that allow the complete and unambiguous positive identification of NK cells in either mouse or man. For example, in the mouse, the markers most commonly used are asialoGM1 and NK1.1. These function reasonably well (albeit imperfectly, being expressed on some macrophages and, as discussed above, on rare populations of T cells) for identifying NK cells in fresh lymphocyte preparations, and can be helpful for purification, but are useless for cultured cells, as most if not all of the T cells that can grow in IL-2 either initially express or rapidly acquire these markers (Brooks *et al.*, 1983). A further problem with NK1.1 is that it is only expressed in certain strains of mice, principally C57 strains. Two promising new antibodies have recently become available. DX5 is highly specific for NK cells and has the major advantage over anti-NK1.1 that it reacts with most or all NK cells in all strains of mice tested. However, it also reacts with a minor subpopulation of T cells (presumably NK-T cells), and has a major disadvantage from the point of view of culturing NK cells in that its expression on cultured NK cells is very low (C. G. Brooks, unpublished). 10A7 reacts with about 70% of NK cells but unlike NK1.1 appears to be absent from the vast majority of T cells, even after prolonged culture in IL-2 (C. G. Brooks and V. Kumar, unpublished). The disadvantage with this antibody is that, like NK1.1, the relevant antigen is only expressed in certain strains of mice, such as C57. In addition, the mAb is currently available only from the original investigator (Professor V. Kumar, Dallas, USA).

Similar problems exist in the human where, although several antibodies have been described that react preferentially with NK cells, none of them is absolutely specific. The antibodies most commonly used for identifying human NK cells are anti-CD16 (FcγRIIIA) and anti-CD56 (NCAM). CD56 is considered to be present on all NK cells, whereas CD16 is found at high levels on around 95% of the NK cells in adult peripheral blood, the remaining CD56$^+$ cells having no or only low levels of CD16 (Nagler et al., 1989). The developmental relationship between these cells is unclear. Other markers present on all human NK cells are NKRP-1, recognised by the antibody DX1, and CD94.

Since the major problem in identifying and purifying NK cells is distinguishing them from T cells, the most satisfactory solution is to combine the presence of an 'NK-specific' marker with the absence of the T cell-specific marker CD3ε, at least at the cell surface. Thus, for identification purposes the simplest procedure is to perform two-colour immunofluorescence/flow cytometry with appropriate antibodies. For cell separation, such stained populations can be purified by cell sorting, but where such facilities are not available or where the frequency of NK cells is low, alternative cell separation procedures must be employed. A simplifying aspect from the point of view of the culturing of NK cells is that, once T cells have been excluded, the only cells present in mouse spleen or human blood that will grow in IL-2 alone are NK cells, so the culture procedure itself can fulfil a critical role in the overall process of purification.

General materials and equipment needed for the culture of NK cells

Equipment

A sterile cabinet (class II cabinet for human work, horizontal laminar flow cabinet for mouse work).

Plasticware and glassware

13–15 ml conical-bottomed centrifuge tubes
30–50 ml conical-bottomed Universal or centrifuge tubes
1.5 ml microfuge tubes
24-well culture plates

25 cm^2 and 75 cm^2 culture flasks
Bacteriological (i.e. not tissue culture grade) 90 mm Petri dishes
Disposable glass or plastic Pasteur pipettes
Disposable micropipette tips.

In our experience products from any of the major suppliers are suitable. We generally use recycled washed glass pipettes, but for the most critical procedures (e.g. cloning of mouse NK cells) the use of disposable plastic pipettes is preferable. Washed glass pipettes and disposable glass Pasteurs are sterilised in metal cans in a 180 °C oven for 2 h. Plastic tips are sterilised in a 120 °C oven for 3 h. Autoclaving should be avoided as it can result in the deposition of toxic material on surfaces.

Culture media

Hanks' balanced salt solution (HBSS)
Dulbecco's minimal essential medium (DMEM)
Yssel's medium (see Chapter 4)
Non-essential amino acids
Fetal bovine serum (FBS)
Human serum
2-mercaptoethanol

Preparation of culture medium

For the long-term growth and cloning of mouse NK cells we have found that supplies of commercial 1 × liquid medium are often inferior to medium made up from powdered medium and highly purified water (we used glass–distilled water that has been passed through a Nanopure deionizer (Whatman, Maidstone, UK) and subsequently stored at 4° in scrupulously cleaned glass bottles with liner-free caps). For washing and preparing mouse cells we use HBSS lacking bicarbonate and supplemented with 1, 2, 5, or 10% FBS (giving H1F, H2F, H5F, and H10F). The HBSS is made up from powder (Life Technologies, Paisley, U.K., 61200-093) in purified water and filter sterilised. For mouse cell culture we use high-glucose DMEM (52100-039) supplemented with 2 × non-essential amino acids (11140-035), 5 × 10^{-5} M 2-mercaptoethanol, and 10% FBS. For FBS we use Sigma F7554 and have never found any need to batch test. FBS should be stored at −80 °C until needed, and should *not* be heated at 56 °C. For culture of human NK cells we use Yssel's medium supplemented with 1% pooled human serum. The composition of this medium and method of preparation have been described in Chapter 4 of this volume. To minimise the risk of mycoplasma infection, antibiotics should be omitted from culture media.

Identification of NK cells by immunofluorescence and flow cytometry

Antibodies

FITC PK136 anti-NK1.1
FITC DX5
PE 2C11 anti-mouse CD3ϵ
FITC anti-human CD56
FITC anti-human CD94
PE anti-human CD3
2.4G2 anti-mouse CD16/32 (Fc block)

For unambiguous identification of NK cells in the mouse, cells should be double stained with FITC PK136 anti-NK1.1 (for C57 strains) or FITC DX5 (for other strains) together with PE 2C11 anti-CD3ϵ (antibodies can be purchased from Pharmingen, San Diego, USA and some other companies). For human NK cells we would recommend the use of FITC anti-CD56 from Becton-Dickinson, San José, USA, FITC anti-CD94 from Pharmingen, or PE anti-CD94 from Immunotech, together with a PE anti-CD3 antibody (many suppliers). These and other antibodies can be stored frozen in small aliquots (25–100 µl) indefinitely at −80 °C. Working solutions can be stored at 4 °C for many months at appropriate dilution in HBSS containing 2% FBS and 0.2% sodium azide (H2FA). The antibodies should be titrated to determine the dose that gives adequate staining of NK cells or T cells with little or no background staining of irrelevant cells (in the mouse the latter can sometimes be reduced by using mAb 2.4G2 (from Pharmingen) to block Fc receptors).

Typical protocol

1 Place 25 µl aliquots of FITC anti-NK and PE anti-CD3 mAbs at 3× the desired concentration in a 1 ml microfuge tube. Negative control tubes should be set up with 50 µl medium and single-stained positive controls with 25 µl of medium and 25 µl of each mAb alone.

2 Add 25 μl containing 1–5×10⁵ cells to each tube, mix well, and place in ice for 20 min.

3 Add 1 ml of H2FA, mix, spin in a microfuge at 10–15000g for 20 s.

4 Aspirate supernatant, add 0.5 ml H2FA, resuspend cells by vortexing.

5 Run cells on a flow cytometer, following the operating procedures described in the manufacturer's manual. Use a forward scatter threshold to exclude debris, and for double-stained samples, FITC/PE compensation should be set carefully, if necessary by making artificial mixtures of unstained and single-fluorochrome-stained cells. (Where the frequency of NK cells or T cells is low a separate appropriately stained sample will be needed – for mouse cells we use fresh spleen cells double stained with FITC goat anti-mouse Ig and PE anti-CD3.)

6 Collect 10–50×10³ cells and analyse on a FITC versus PE dot plot having gated tightly on viable single-nucleated cells using forward and side-scatter parameters. The percentages of NK cells (FITC⁺PE⁻), conventional T cells (FITC⁻PE⁺), NK-T cells (FITC⁺PE⁺), and other contaminants (FITC⁻PE⁻) can be readily determined using quadrant markers.

Culture of NK cells

Because at least some NK cells proliferate autonomously (without any requirement for accessory or feeder cells) and directly (without any requirement for a stimulus) in the presence of suitable concentrations of IL-2, short-term culture of these cells is in principle a simple matter. However, as indicated above, because various types of T cell will proliferate under the same conditions, obtaining pure NK cell cultures is a difficult and demanding procedure. The matter is compounded by the fact that the offending T cells often proliferate more rapidly than the NK cells so that even an initial minor contamination can result in cultures being overgrown by T cells within a week or so. A further problem is that in both mouse and human the proliferative capacity of NK cells obtained from spleen or blood is finite, and after a few weeks of rapid proliferation growth ceases. Although NK cells can proliferate in response to a number of lymphokines including IL-4, IL-7, IL-12, IL-15, and TNF, in our experience none of these alternative growth factors either alone or in combination provides any advantage over IL-2, and they have the disadvantage of being harder to obtain and being much more expensive. As in the case of T cells, human IL-2 is as effective in promoting the growth of mouse NK cells as is mouse IL-2 (C. G. Brooks, unpublished).

Purification and culture of mouse NK cells

The optimal procedure is to first deplete conventional T cells with antibody plus complement (Ab + C′), then to remove B cells by panning, followed by removal of red cells and dead cells by density step centrifugation, and removal of residual T cells by further panning, magnetic beads, or cell sorting.

Materials

Anti-mouse CD4
Anti-mouse CD8

These reagents should be purchased in sterile form or filter sterilised on receipt. Complement-fixing mouse or rat mAbs should be chosen, preferably of the IgM class (we use the rat IgM mAbs RL172.4 and 3.158 available from the American Type Culture Collection, Rockville, USA). The concentrations required to give effective depletion of T cells should be determined in preliminary experiments using the protocol described below, followed by staining with PE anti-CD3 (see above) or (preferably, as it is more sensitive) by measuring responses of the treated spleen cells to ConA and LPS.

Rabbit C′

We prepare our own as serum from young outbred rabbits and store it frozen in suitable aliquots at 4 °C. Such serum often contains natural anti-mouse T cell antibodies, which can aid the process of purification of NK cells, and is often potent enough to be used at a high dilution (eg. 1/40). A more expensive alternative is to use commercial rabbit C′ (eg. Lowtox M C′ from Cedarlane, Hornby, Canada, used at 1/10 or higher concentration). The optimal concentration of C′ should be determined in conjunction with titration of the anti-CD4 and anti-CD8 mAbs in a 'checkerboard' experiment.

Affinity-purified anti-mouse Ig

This reagent is available from many commercial sources (we generally use antibody made in sheep or goats as this (supposedly) does not bind to mouse Fc receptors). To coat flasks for panning, prepare an appropriate volume (1.5 ml for 25 cm² flasks, 4.5 ml for 75 cm² flasks) of Ab at 10 μg/ml in borate-buffered saline and add immediately to flasks, dispersing it evenly and completely over the surface to be used for panning. Incubate in the flasks overnight at room temperature. Just before panning, aspirate the remaining solution, and wash the flasks thoroughly with PBS or HBSS (no serum).

Lympholyte M (Cedarlane)

KT3 anti-mouse CD3e (Serotec, Kidlington, Oxford, UK)

The concentration required to give maximal staining of mouse T cells should be determined by indirect immunofluorescence/flow cytometry.

Magnetic beads coated with sheep anti-rat Ig (Dynal, Oslo, Norway)

Magnet for bead separation (Dynal)

Recombinant human IL-2

This is available from a multitude of commercial sources. Given the large amounts of IL-2 required, costs can be substantial, so it is worth shopping around and trying to negotiate a discount. In our experience the functional IL-2 receptors on mouse NK cells are not saturable at least up to 10^5 units/ml. Thus the more IL-2 is used the greater the growth rate and yield of NK cells. We routinely use IL-2 at 10^4 U/ml, but some growth can be achieved with 10^3 U/ml. We measure units against the WHO international IL-2 standard using the CTLL2 bioassay (Gillis et al., 1978).

Protocol

1 Remove spleen(s) from freshly killed mice into a Petri dish containing 10 ml H1F (see Preparation of culture medium). Make a few small incisions in the spleen(s) with scissors or scalpel blade, then tease out the splenocytes using large forceps. Avoiding as much as possible residual splenic tissue, make a uniform suspension of splenocytes by pipetting the suspension up and down several times in a 10 ml pipette, and transfer to a 13–15 ml centrifuge tube. If using more than one spleen, divide the suspension between several tubes at the rate of one spleen/tube, and make up the volume of each to 10 ml with H1F.

2 Spin cells at 400g for 5 min at room temperature, aspirate supernatant, disperse the pellet by tapping the tube vigorously against a hard surface (this is crucial to achieve uniform resuspension of cells, and should be used after each spinning of cells), and resuspend in 10 ml H1F.

3 Allow the tube to stand for 10-20 min to allow tissue clumps and cell aggregates to settle, and transfer the clump-free supernatant to a new tube. While the clumps are settling the concentration of viable white cells should be determined by counting in a haemocytometer.

4 Spin cells at $400g$ for 5 min at room temperature, and resuspend in H5F so as to give 40×10^6 cells/ml. Add anti-CD4 and anti-CD8 mAbs from working solutions stored at a $4 \times$ concentration in H5F so that the final cell concentration is 20×10^6/ml. Do not have more than 150×10^6 cells/tube. Mix by inverting the tube several times, and place in ice for 30 min.

5 Spin at $400g$ for 5 min at room temperature, and resuspend in freshly thawed rabbit C' (diluted to the predetermined optimal concentration in H5F) so that the cells are again at a nominal 20×10^6/ml. Mix by inversion, and transfer the suspension to a new tube. Place the tube in a 37 °C water bath for 45 min, making sure it is immersed up to the level of the cell suspension.

6 Spin at $400g$ for 5 min, and resuspend in H5F. Count and adjust to 4×10^6 viable white cells/ml.

7 Perform a first round of panning by incubating cells in plain uncoated 75 cm² tissue culture flasks incubated horizontally and left undisturbed for 1 h at room temperature at the rate of 5 ml per small (25 cm²) flask or 15 ml per large (75 cm²) flask. We have discovered that this simple procedure is as effective as nylon-wool columns for removing B cells.

8 Recover non-adherent cells by gently rocking the flask(s) and transfer to anti-mouse Ig-coated flasks, then incubate as above.

9 Recover the non-adherent cells, spin down in 13–15 ml centrifuge tubes at $1000g$ for 10 min at room temperature. Resuspend in a total of 3 ml H5F. The two panning procedures will have resulted in a loss of >90% of white cells, mostly at the first round.

10 Underlay with ~1 ml of Lympholyte M (this should be brought to room temperature before use). Spin at $1000g$ for 20 min with slow acceleration and no brake. This removes the now large excess of red cells and dead cells, and also (contrary to common knowledge) some residual T cells and B cells, leading to a fortuitous slight enrichment of NK cells.

11 Collect the cells from the interface with a Pasteur pipette, in a volume of about 1 ml. Add 4 ml HIF and spin at $1000g$ for 10 min at room temperature.

12 Resuspend the cells in ~100 µl of H2FA and add 100 µl of KT3 anti-CD3 mAb to give the predetermined saturating concentration. Mix, and incubate on ice for 20 min.

13 Wash the cells twice in the cold with 3 ml of H2FA, spinning at 1000*g* for 5 min at 4°C (make sure centrifuge has been precooled to 4°C). Resuspend the cells finally in 3 ml H2FA and count.

14 Place pre-washed (see manufacturer's instructions) anti-rat Ig-coated Dynabeads in a second tube, ten Dynabeads for every cell. Dilute these with 3 ml HIF, and spin down the tube of cells and the tube of beads in the cold.

15 Resuspend the cells in 200 μl of H5F, add these to the pellet of beads in the second tube, and mix, taking care to avoid any bubbling. Incubate at room temperature for 15 min, *gently* mixing the tube every minute.

16 Spin the cells down at very slow speed (100*g*) for 2 min with no brake. Without removing the supernatant, *gently* resuspend the cells with a pipetter and dilute to 5 ml with H1F. Mix and stand the tube against the magnet for 2 min. Without moving the tube carefully transfer the bead-free suspension to a second tube and repeat the magnet treatment. Spin the cells down and resuspend in culture medium.

By this stage only about 1% of the starting cells will be left. Typically these will comprise 60-80% NK cells, contaminated with a few residual B cells and miscellaneous uncharacterised cells. The cells should be placed in a single well of a 24-well plate in 2 ml D10F containing 10^4 U/ml IL-2. Within 2–3 days, the well will contain dense clusters of vigorously proliferating cells, and these should be re-fed and/or subcultured as needed every 3–4 days. Mouse NK cells proliferate rapidly for about 2 weeks, then more slowly for a further 1–2 weeks, before entering a static stage in which viable cell numbers gradually decline. If re-feeding is continued such static NK cell cultures can sometimes be maintained in a healthy and functional (in terms of cytolytic activity) state for several weeks, but for unknown reasons the cells often transform into very large granular cells, which are of no use. If any T cells remain after the purification procedure these will gradually become the predominant cell type in such cultures. Thus, checking the purity of NK cell cultures as described above is crucial.

Cloning of mouse NK cells

The limited proliferative potential of normal adult mouse NK cells has provided a major barrier to the cloning of these cells. Two recent approaches, however, have offered at least a partial solution to this problem. One approach, followed by Karlhofer, Orihuela & Yokoyama (1995) has been to use mice in which the tumour-suppressor p53 gene has been functionally inactivated. Lymphoid cells from such mice appear to have unlimited pro-liferative capacity *in vitro*, allowing long-term proliferating NK cell lines with an apparently indefinite life span to be obtained, from which clones can be obtained in the presence of appropriate feeder cells. An alternative, and potentially more general, approach exploits the fact that fetal cells have a greater proliferative capacity than adult cells and that, somewhat surprisingly, the early fetal thymus turns out to be a rich source of clonogenic NK cell precursors (Brooks, Georgiou & Jordan, 1993). Under appropriate condi-tions the majority of individual day 14 fetal thymocytes can develop into NK cell clones *in vitro*, and with care a large proportion of these can be grown for long periods. The fetal system has the further advantages that no enrich-ment or purification of NK cells is required, and clones develop vigorously in the absence of any feeder cells. Thus, cloning can be performed if desired in the most rigorous way possible, i.e. by micromanipulation, and clonal development can be visualised and studied from the very earliest stages. The key to success in this system was the discovery that brief exposure of day 14 fetal thymocytes to IL-4 plus PMA induces a greatly enhanced responsive-ness to IL-2 (Brooks *et al.*, 1993). Long-lived NK cells lines of apparently different maturational status can also be obtained from fetal liver using the same approach, although we have had less success in generating clones from fetal liver (Manoussaka *et al.*, 1997). Curiously, NK cell lines and clones gen-erated in both the p53−/− system and from fetal precursors are generally deficient in expression of Ly49 molecules. However, in all other respects they have proven to be indistinguishable from short-term cultured adult splenic NK cells.

Materials

Timed-mated mice

These can either be set up in-house or purchased commercially, for example from Bantin & Kingman, Hull, UK. They should be used at day 14, where the day of mating/plug detection is counted as day 0.

Recombinant mouse IL-4

Purified recombinant mouse IL-4 can be purchased from many companies. The dosage of this material is crucial, and should be determined in prelim-inary experiments by titrating for induction of proliferation in fetal thymo-cytes in the presence of PMA, 1 unit/ml being the concentration of IL-4 that gives 50% maximal proliferation.

Phorbol myristate acetate (PMA)

Purchased from Sigma. PMA is highly unstable in aqueous solution. The vial should be reconstituted with pure ethanol to give a concentration of 1 mg/ml, then can be stored at −20 °C indefinitely. Just before use a working aqueous solution should be prepared at 1 μg/ml.

96-well half-area cloning plates (Corning/Costar, Acton, USA)

These are considerably more expensive than regular 96-well plates, but make the process of screening for clones much faster.

Protocol

1 Remove thymus lobes from the fetuses of day 14 timed-mated mice, placing them in ice-cold H10F. This is an intricate procedure (for details see Chapter 3), which requires a dissecting microscope and some specialist training. One average litter's worth of thymuses would provide enough cells (typically $20\text{-}50 \times 10^3$ cells per lobe).
2 Using cataract knives, tease the thymuses in a droplet of about 0.5 ml of H10F in a Petri dish. Rinse the cells into a tube using about 2 ml of H10F, and allow the clumps to settle for 10-20 min before transferring the clump-free supernatant to a fresh tube.
3 Spin at 1000g for 5 min, and resuspend in 1 ml D10F. Perform a cell count. Culture up to 1×10^6 cells in a well of a 24-well plate in D10F containing 10 U/ml IL-4 and 10 ng/ml PMA.
4 After 24–48 h, by which time it should be clear that the cells are engaged in rapid proliferation, count the cells and clone them by limiting dilution or micromanipulation in IL-2 plus PMA.
5 After 7, 10, and 14 days of incubation, plates should be examined for the presence of developing colonies, and these should be transferred to the wells of 24-well plates and maintained by refeeding and/or subculturing twice a week with the same medium. Only clones with a high probability of monoclonality (>95%) as calculated from Poisson statistics should be retained.
6 After about 1 month of growth, foetal NK cell clones enter a phase of slow growth. However, if looked after carefully by refeeding at least twice per week with medium containing 10^4 U/ml IL-2 and PMA, with occasional subculture most clones resume moderate to rapid growth, and can usually then be maintained indefinitely as continuously proliferating lines. We

have found that the success rate in getting clones through this slow growing stage is enhanced if small amounts of IL-4 (around 0.5 U/ml) are included in the medium. Use of higher amounts of IL-4 is counter-productive, as it can either inhibit growth or induce the cells to differentiate into very large and very granular cells, which eventually die out.

Purification and culture of human NK cells

Materials

Medium: For culture of human NK cells we use Yssel's medium (Yssel *et al.*, 1984) supplemented with 1% pooled human serum. The composition of this medium and method for preparation have been described in Chapter 4 by Spits and Brooks.

Purification

Probably the most direct way to enrich NK cells is by centrifugation over a Percoll gradient. The method was originally described by Timonen & Saksela (1980). We use a modification introduced by Phillips & Lanier (1986). This method takes advantage of the higher density of T cells compared to NK cells. Since monocytes are also of a light density, these cells are removed from PBMC by adherence on plastic for 45 min at 37 °C. This is followed by passage through a nylon-wool column (various suppliers, eg. Cellular Products Inc., Buffalo, New York), which removes residual monocytes and B cells. Make up 30% and 40% Percoll (Pharmacia Biotech, St. Albans, UK) in RPMI-1640 supplemented with 10% FCS. The cells are resuspended in 30% Percoll, which is carefully layered over 40% Percoll. Following 30 min centrifugation at 1000*g* in a table-top centrifuge (with brakes off), the interface is enriched for large granular lymphocytes, which include NK cells and large T cell blasts, while the high density cells, mostly T cells, are on the bottom. The interface is carefully removed and washed three times. This population should contain at least 50% NK cells. Residual T cells may be removed by magnetic bead sorting using anti-CD3 beads (Dynal) and methods similar to those described above for mouse NK cells to obtain a reasonably enriched population (>90% NK cells).

Culture

Mature human NK cells proliferate most efficiently in IL-2 or IL-15. As the receptors for both factors share the IL-2Rβ and γ chains, these cytokines presumably induce growth of NK cells in the same way. IL-4 and IL-7 have been reported to sustain growth of NK cells but in our hands IL-4 does not induce growth of human NK cells while IL-7 is very inefficient compared to IL-2. Hence, for reasons of simplicity, rIL-2 is used in our laboratory to culture NK cells. Human peripheral blood NK cells grow rather poorly in IL-2 alone. $CD56^{++}CD16^-$ NK cells proliferate to a reasonable extent. However, $CD56^+CD16^{++}$ NK cells can be activated by IL-2 to produce IFNγ but proliferate hardly at all. This is consistent with the finding that resting $CD16^-$ NK cells express both IL-2 receptor (R) α and β chains, while resting $CD16^+$ NK cells express IL-2Rβ but no IL-2Rα chain (Caligiuri *et al.*, 1990; Nagler, Lanier & Phillips, 1990). Human NK cells grow much better in the presence of a feeder cell mixture and PHA. We use the same feeder cell mixture (i.e. irradiated allogeneic PBMC and irradiated EBV-transformed B cells) and conditions for growing human NK cells as described in Chapter 4 for growing human T cells. Different subsets of human NK cells from various organs of both fetus and adult can be cultured using this method.

Since both T and NK cells grow under these conditions, T cells need to be rigorously depleted before culturing the NK cells, as described above. Depletion of T cells using magnetic beads with T cell-specific antibodies (anti-CD3 and anti-CD5) is often not sufficient to obtain adequately pure NK cells. The few contaminating T cells may rapidly overgrow the NK cells. To ensure a complete removal of T cells, FACS sorting is required. If such equipment is not available, one should regularly check the NK cultures for the presence of T cells by immunofluorescence staining and flow cytometry as described above, followed by removal of the T cells using magnetic beads, cell sorting or other suitable methods.

Cloning of human NK cells

Human NK cells can be cloned either by limiting dilution or by micromanipulation. If the equipment is available, micromanipulation can be carried out by single-cell FACS cloning. The procedures are exactly as those describe in Chapter 4 for human T cells. Cloning efficiencies of the two subsets of mature NK cells ($CD16^{++}$ and $CD16^{-/+}$ cells) from adult PBMC are similar and usually around 2–8% (J. H. Phillips & H. Spits, unpublished) which is low compared to those of subsets of PBMC T cells (20–100%). However, addition of rIL-2 at the outset of cloning resulted in a higher cloning efficiency (10–40%) of $CD16^-$ NK cells, while immediate addition of IL-2 had no effect on the cloning efficiencies of $CD16^+$ NK cells or of T cells.

References

Ballas, Z. K. & Rasmussen, W. (1990). Lymphokine-activated killer (LAK) cells. IV Characterization of murine LAK effector subpopulations. *J. Immunol.*, **144**, 386–95.

Brooks, C. G. (1983). Reversible induction of natural killer cell activity in cloned murine cytotoxic T lymphocytes. *Nature*, **305**, 155–8.

Brooks, C. G., Burton, R. C., Pollack, S. B. & Henney, C. S. (1983). The presence of NK alloantigens on cloned cytotoxic T lymphocytes. *J. Immunol.*, **131**, 1391–5.

Brooks, C. G., Georgiou, A. & Jordan, R. K. (1993). The majority of immature foetal thymocytes can be induced to proliferate to IL-2 and differentiate into cells indistinguishable from mature NK cells. *J. Immunol.*, **151**, 6645–56.

Caligiuri, M. A., Zmuidzinas, A., Manley, T. J., Levine, H., Smith, K. A. & Ritz, J. (1990). Functional consequences of interleukin 2 receptor expression on resting human lymphocytes. Identification of a novel natural killer cell subset with high affinity receptors. *J. Exp. Med.*, **171**, 1509–26.

Gillis, S., Ferm, M. M., Ou, W. & Smith, K. A. (1978). T cell growth factor: parameters of production and a quantitative microassay for activity. *J. Immunol.*, **120**, 2027.

Herberman, R. B., Balch, R., Bolhuis, R., Golub, S., Hiserodt, J., Lanier L., Lotzova, E., Phillips, J., Riccardi, C., Ritz, J., Santoni, A., Schmidt, R., Uchide, A. & Vujanovic, N. (1987). Most lymphokine activated killer (LAK) activity mediated by blood and splenic lymphocytes is attributable to stimulation of natural killer (NK) cells by interleukin-2 [IL-2]. *Immunol. Today*, **8**, 178–84.

Hori, T., Phillips, J. H., Duncan, B., Lanier, L. L. & Spits, H. (1992). Human foetal liver-derived CD7$^+$CD2lowCD3$^-$CD56$^-$ clones that express CD3 gamma, delta, and epsilon and proliferate in response to interleukin-2 (IL-2), IL-3, IL-4, or IL-7: implications for the relationship between T and natural killer cells. *Blood*, **80**, 1270–8.

Hori, T. & Spits, H. (1991). Clonal analysis of human CD4$^-$CD8$^-$CD3$^-$ thymocytes purified from postnatal thymus. *J. Immunol.*, **146**, 2116–21.

Karlhofer, F. M., Orihuela, M. M. & Yokoyama, W. M. (1995). Ly49-independent natural killer (NK) cell specificity revealed by NK cell clones derived from p53-deficient mice. *J. Exp. Med.*, **181**, 1785–95.

Koyasu, S. (1994). CD3$^+$ CD16$^+$ NK1.1$^+$ B220$^+$ large granular lymphocytes arise from both α-βTCR$^+$ CD4$^-$ CD8$^-$ and γ-δTCR$^+$ CD4$^-$ CD8$^-$ cells. *J. Exp. Med.*, **179**, 1957–72.

Linnemeyer, P. A. and Pollack, S. (1993). Prostaglandin E2-induced changes in the phenotype, morphology and lytic activity of IL-2 activated natural killer cells. *J. Immunol.*, **150**, 3747–54.

Manoussaka, M. M., Georgiou, A., Rossiter, B., Shrestha, S., Toomey, J. A., Sivakumar, P. V., Bennett, M., Kumar, V. & Brooks, C. G. (1997). Phenotypic and functional characterisation of long-lived NK cell lines of different maturational status obtained from mouse foetal liver. *J. Immunol.*, **158**, 112–19.

Nagler, A., Lanier, L. L., Cwirla, S. & Phillips, J. H. (1989). Comparative studies of human FcRIII-positive and negative NK cells. *J. Immunol.*, **143**, 3183–91.

Nagler, A., Lanier, L. L. & Phillips, J. H. (1990). Constitutive expression of high affinity interleukin-2 receptors on human NK cell subsets. *J. Exp. Med.*, **171**, 1527–33.

Pantaleo, G., Zocchi, M. R., Ferrini, S., Poggi, A., Tambuni, G., Bottino, C., Morreta, L. & Moretta, A. (1988). Human cytolytic cell clones lacking surface expression of T cell receptor $\alpha\beta$ or $\gamma\delta$. Evidence that surface structures other than CD3 or CD2 molecules are required for signal transduction. *J. Exp. Med.*, **168**, 13–24.

Pawelec, G. P., Hadam, M. R., Ziegler, A., Lohmeyer, J., Rehbein, A., Kumbier, I. & Wernet, P. (1982). Long-term culture, cloning and surface markers of mixed leukocyte culture-derived human T lymphocytes with natural killer-like cytotoxicity. *J. Immunol.*, **128**, 1892–6.

Patel, S. S., Wacholtz, M. C., Duby, A. D., Thiele, D. L. & Lipsky, P. E. (1989). Analysis of the functional capabilities of CD3[+] CD4[-] CD8[-] and CD3[+] CD4[+] CD8[+] human T cell clones. *J. Immunol.*, **143**, 1108–17.

Phillips, J. H. & Lanier, L. L., (1986). Dissection of the lymphokine-activated killer phenomenon: relative contribution of peripheral blood natural killer cells and T lymphocytes to cytolysis. *J. Exp. Med.*, **164**, 814–25.

Shortman, K., Wilson, A. & Scollay, R. (1984). Loss of specificity in cytolytic T lymphocyte clones obtained by limit dilution of Ly2[+] T cells. *J. Immunol.*, **132**, 584–92.

Spits, H., Lanier, L. & Phillips, J. H. (1995). Development of human T and natural killer cells. *Blood*, **85**, 2654–70.

Timonen, T. & Saksela, E. (1980). Isolation of human NK cells by density gradient centrifugation. *J. Immunol. Methods*, **36**, 285–91.

Yssel, H., De Vries, J. E., Koken, M., van Blitterswijk, W. & Spits, H. (1984). Serum-free medium for the generation and the propagation of functional human cytotoxic and helper T cell clones. *J. Immunol. Methods*, **72**, 219–27.

8

Mast cells and basophils

Gunnar Nilsson and Dean D. Metcalfe

Introduction and application

Mast cells and basophils are the primary effector cells in immediate hyper-sensitivity reactions. Both originate from pluripotential haematopoietic cells and share the unique characteristics of expression of the high affinity receptor for IgE on their surface and the ability to synthesise and store histamine and cytokines. Mast cells are normally found exclusively in tissue, while basophils are found in the blood, where they normally represent less than 3% of the leukocytes. Among mast cells there exists a heterogeneity which in humans is determined on the basis of the protease content. Mast cells are thus referred to as having a MC_T (mast cell which is tryptase positive, chymase negative) or MC_{TC} (for tryptase positive, chymase positive) phenotype.

Much of our current knowledge concerning the regulation of mast cell and basophil differentiation has resulted from advances in techniques for culturing these cells *in vitro*. Until the identification of IL-3 as a rodent mast cell growth factor in the early 1980s, systems to culture mast cells were difficult and inadequate. However, due to species differences, it was not until the identification of stem cell factor (SCF), also referred to as kit-ligand, steel factor, or mast cell growth factor (see Galli, Zsebo & Geissler, 1994 and references therein) that it was possible to culture human mast cells *in vitro*. In contrast to that response in rodents, IL-3 does not induce any significant differentiation of human mast cells, although it does support human basophils.

The identification of IL-3 as a growth factor for human basophils and SCF for human mast cells has allowed the study of the growth and differentiation of both human basophils and mast cells and examination of their biological potential in activation, adhesion, migration, and receptor expression assays. In this chapter we will review the culture techniques for both mouse and human mast cells and basophils.

Protocol to culture murine mast cells and basophils

Murine mast cells develop in bone marrow cultures supplemented with recombinant IL-3 (rIL-3), rSCF or conditioned medium containing IL-3 (see below). Murine mast cells grown in IL-3 alone are immature. Addition of SCF promotes mast cell maturation as evidenced by histochemistry and by the expression of a distinct granule protease phenotype (Gurish *et al.*, 1992). Other growth factors and cytokines, such as nerve growth factor, IL-9 and IL-10, also influence differentiation and alter the granule protease phenotype (Eklund *et al.*, 1993).

There is no technique for the exclusive culture of murine basophils. A significant number of basophils are detected up to three weeks in bone marrow cultured in IL-3 with or without SCF (Rottem *et al.*, 1993). Cytospin preparations of cultured mast cells and basophils are shown in Fig. 8.1.

Materials and equipment

Surgical tools, several sets of sterile forceps, scissors and containers
70% ethanol
Petri dishes (10 cm in diameter)
50 ml conical polypropylene tubes
syringes (5 ml)
needles (gauge 23)

Medium

RPMI 1640 supplemented with:

10% heat inactivated fetal calf serum (FCS)
10 ng/ml rIL-3 or 10–20% WEHI-3 conditioned medium (see below)
25 mM HEPES
4 mM L-glutamine
1 mM sodium pyruvate
0.1 mM non-essential amino acids
50 µM mercaptoethanol
100 µg/ml penicillin/streptomycin

All ingredients can be supplied from Sigma Chemical Co., St. Louis, MO; Life Technologies, Gaitersburg, MD or a similar company.

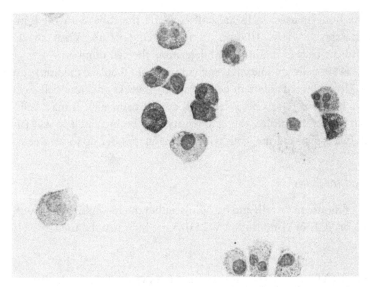

Fig. 8.1. Morphology of SCF-dependent mast cells cultured from CD34$^+$ bone-marrow-derived cells. The cells are stained with toluidine blue. Courtesy of Dr Arnold Kirshenbaum.

Procedure

1 Animals: 6- to 9-week-old mice (BALB/c). Bone marrow from one to five mice are pooled.
2 The mouse is killed by cervical dislocation. The mouse may first be anaesthetised with Metofane.
3 Soak the area of incision with 70% ethanol to reduce the risk of contaminating the dissection with mouse hair.
4 Place the animal on its side, pinch the skin and make an incision on the thigh. Dissect the skin to expose the muscles. Dislocate the femur or cut free the femur above the hip joint and below the stifle joint. Place the femur unseparated from muscles in a sterile Petri dish with medium.
5 Continue the procedure in a cell culture hood. Use a second set of scissors and forceps to separate the bone from muscle. Transfer the bones to another Petri dish filled with medium and keep covered until the next step.
6 Hold the bone with forceps and cut it on the proximal and distal ends to expose the marrow. Take a syringe with warm medium and insert the needle 2–3 mm into the marrow cavity. Slowly flush the bone marrow into a conical tube.

7 When the bone marrow cells are collected, fill the tube with medium and centrifuge at 400g, 10 min, at room temperature. Wash twice. Resuspend the cells in medium and determine the cell number.

8 Place 1–2 × 10^5 cells/ml in each tissue culture flask. Replace medium and flask weekly to discard adherent cells. After 3–4 weeks adherent cells will no longer be seen. Remaining cells will consist primarily of mast cells. After four weeks in culture, the cell density may be increased to 4 × 10^5 cells/ml. In a typical culture, cells will increase in number up to six weeks.

Conditioned medium

For culture of mouse mast cells and basophils, either the recombinant growth factors IL-3 or SCF, or conditioned WEHI-3 medium may be used.

WEHI-3

The WEHI-3 (ATCC TIB-68) is a myelomonocytic cell line which produces IL-3. The cells are grown in RPMI 1640 medium supplemented with 10% heat inactivated FCS, 25 mM HEPES, 4 mM L-glutamine, 100 µg/ml penicillin/streptomycin, 0.1 mM nonessential amino acids and 1 mM sodium pyruvate. Cells are cultured at 2–3 × 10^5 cells/ml. Conditioned medium is collected every 3 to 4 days, filtered through a 0.22 µm filter, and stored at −20 °C until use. WEHI-3 conditioned medium (10–20%) is used to culture mouse basophils and mast cells.

Protocols to culture human mast cells and basophils

Human peripheral blood, cord blood, or bone-marrow-derived mononuclear cells separated by density sedimentation and cultured in IL-3-containing media yield cultures where basophils predominate at 2–3 weeks. If SCF is used instead of IL-3, an almost pure population of mast cells will be obtained after 8–10 weeks of culture. Addition of IL-6 to these cultures further promotes the SCF-dependent development of human mast cells (Saito et al., 1996). No protocols are available that yield a specific human mast cell phenotype (MC$_T$ or MC$_{TC}$). However, using fetal liver cells as a source for SCF-dependent human mast cell culture, MC$_T$ predominate (Irani et al., 1992). In contrast, if cord blood cells are used MC$_{TC}$ cells predominate (Nilsson et al., 1996).

Mast cells may be cultured from bone marrow, peripheral blood, umbilical cord blood, or fetal liver mononuclear cell preparations or CD34$^+$ cell sorts in the presence of SCF (Irani et al., 1992; Kirshenbaum et al., 1992; Mitsui et al., 1993; Valent et al., 1992). Similarly, human basophils may be cultured from these sources in the presence of IL-3 (Kirshenbaum *et al.*, 1989; Saito *et al.*, 1988; Valent *et al.*, 1989). Protocols for both approaches are given.

Culture of human mast cells and basophils from mononuclear cells

Either mononuclear cell preparations or CD34$^+$ cell populations may be used to culture mast cells and basophils. For studies on cell differentiation it may be preferable to use CD34$^+$ progenitor cells.

Materials and equipment

Heparinised peripheral blood, cord blood, bone marrow or dispersed fetal
  liver cells
50 ml conical polypropylene tubes
Ficoll-Paque (1.077 density) (Pharmacia Biotech, Uppsala, Sweden)
PBS
RPMI 1640 medium supplemented with 10% heat inactivated, either fetal
  calf, autologous human serum, or human AB$^+$ serum, 25 mM HEPES,
  4mM L-glutamine, 1 mM sodium pyruvate, 0.1 mM non-essential
  amino acids, 50 μM mercaptoethanol and 100 μg/ml penicillin/strep-
  tomycin.
rIL-3 at 10 ng/ml for the growth and differentiation of basophils
rSCF at 50 ng/ml for the growth and differentiation of mast cells
rIL-6 at 10 ng/ml for the growth and differentiation of mast cells

Procedure

1 Peripheral blood, cord blood, bone marrow or dispersed fetal liver cell
  preparation is diluted with 2–4 volumes of PBS.
2 Carefully layer 35 ml of the diluted cell suspension over 15 ml of Ficoll-
  Paque in a 50 ml conical tube. Centrifuge at 400g for 30 min at 20 °C in
  a swinging-bucket rotor without braking.
3 Carefully collect the interphase cells and transfer to a new 50 ml conical
  tube. Add PBS to 50 ml and centrifuge at 300g for 10 min at 20 °C.
  Remove the supernatant.
4 Resuspend the cells in PBS and wash twice. Centrifuge at 200g for 10 min
  at 20 °C.
5 Resuspend the cells in complete medium at 1×10^6 cells/ml in tissue
  culture flask or wells. Change medium weekly. For the culture of basophils

add 10 ng/ml of rIL-3 and for culturing mast cells add 50 ng/ml of rSCF and 10 ng/ml rIL-6. The kinetics of appearance of basophils and mast cells in culture is shown in Fig. 8.2.

Growth of human mast cells from CD34-selected progenitor cells

To study the effects of specific growth factors on mast cells or basophil differentiation it may be desirable to enrich for progenitor cells. One marker used for enrichment of haematopoietic pluripotent cells is CD34. Differentiation of mast cells from CD34$^+$ cells occur in the presence of SCF (Kirshenbaum et al., 1992; Saito et al., 1996). An increase in the total number of cells can be obtained by adding IL-3 for the first week of culture.

Antibodies against CD34 are used as tags for enrichment, and cells labelled with antibodies against CD34 may be collected. There are several methods used for this purpose, including panning, flow cytometry and the use of magnetic beads. A protocol using such antibodies and magnetic microbeads, MACS (Miltenyi Biotec Inc., Sunnyvale, CA), is presented below.

Materials and equipment

Ficoll-Paque (1.077 density) (Pharmacia Biotech)
PBS supplemented with 0.5% bovine serum albumin and 2 mM EDTA. Degas buffer by applying vacuum. This prevents the formation of bubbles in the matrix of the column during separation
MiniMACS separation system (Miltenyi Biotec)
CD34 progenitor cell isolation kit (Miltenyi Biotec)
medium as for culture of mononuclear cells
rIL-3, rIL-6 and rSCF

Procedure

1 Peripheral blood, cord blood, bone marrow or dispersed fetal liver cell preparation is diluted with 2–4 volumes of PBS.
2 Carefully layer 35 ml of diluted cell suspension over 15 ml of Ficoll-Paque in a 50 ml conical tube. Centrifuge at 400g for 30 min at 20°C in a swinging-bucket rotor without brake.
3 Collect the interphase cells and transfer to a new 50 ml conical tube. Add PBS to 50 ml and centrifuge at 300g for 10 min at 20°C. Remove the supernatant. Resuspend the cells in 50 ml of PBS and centrifuge at 200g for 10 min at 20°C. Remove the supernatant.

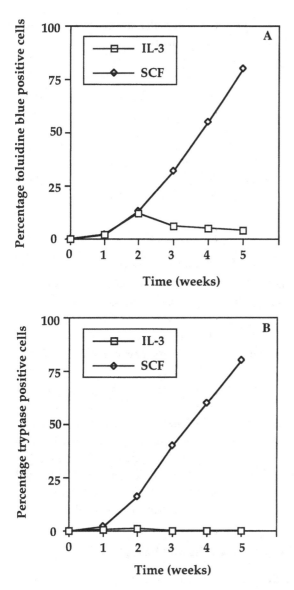

Fig. 8.2. Appearance of toluidine blue positive cells (A) and tryptase positive cells (B) in cultures supplemented with IL-3 or SCF.

4 Resuspend the cells in 50 ml of de-gassed buffer and centrifuge at 200g for 10 min at 20 °C. Remove the supernatant.

5 Resuspend the cell pellet in 300 µl buffer per 10^8 cells.

6 Separate the CD34$^+$ cells with MACS magnetic cell sorting according to the manufacture's protocol. Briefly, add 100 µl of reagent A1 (provided in the cell separation kit) per 10^8 cells and mix well. Add 100 µl of reagent A2, mix well and incubate for 15 min at 6–15 °C.

7 Wash cells with de-gassed buffer by adding 10–20 × the staining volume. Centrifuge at 200g for 10 min at 20 °C and remove supernatant. Resuspend the cell pellet and add buffer to a final volume of 400 µl per 10^8 cells. Add 100 µl Reagent B, mix, and incubate for 15 min at 6–15 °C.

8 Wash cells and resuspend in 500 µl buffer. Pass cells through a 30 µm nylon mesh to remove aggregated cells.

9 Place the column in the magnetic field of the MACS separator. Rinse with buffer. Apply cells to the column, allow non-adhering cells to pass through the column, and wash the column with buffer.

10 Remove column from separator, place column over a tube, and pipette buffer onto the top of column to elute retained cells.

11 For higher purity, repeat steps 9 and 10 with a new column.

12 Wash collected cells with buffer.

13 Resuspend the cells in culture medium (see culture of mononuclear cells). Culture the cells at 1×10^5 cells/ml for the first week in the presence of 50 ng/ml rSCF, 1 ng/ml rIL-3 and 10 ng/ml rIL-6. After one week in culture the cells are grown in 50 ng/ml rSCF and 10 ng/ml rIL-6. Prolonged presence of IL-3 in the cultures will inhibit the differentiation of mast cells (Sillaber et al., 1994). The medium is changed weekly and the cell number is adjusted to 1×10^5 cells/ml. After 8 weeks in culture, more than 90% of the cells in culture are mast cells.

Protocols for detection of mast cells and basophils

The development of basophils or mast cells in cultures is detected either by the use of antibodies directed against cell antigens such as granule-associated proteases and cell surface antigens; by metachromatic dyes that in part bind to highly sulphated proteoglycans within mast cell and basophil granules; or by employing specific substrates for the detection of tryptic enzymes expressed in mast cells (Fig. 8.3). Cytocentrifuge preparations are examined according to the following protocols.

(A)

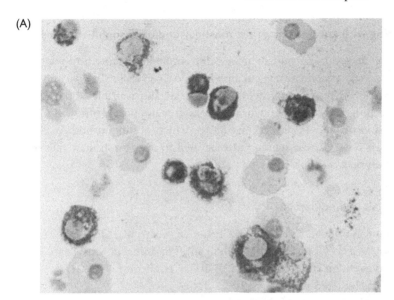

(B)

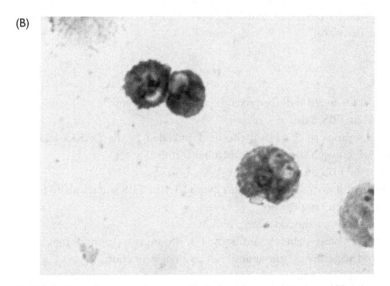

Fig. 8.3. Staining of tryptase in mast cells derived from umbilical cord blood mononuclear cells cultured in the presence of SCF. The cells are stained with an mAb against tryptase (A), or with Z-Gly-Pro-Arg-MNA as the substrate and Fast Black K as the chromogen (B).

Staining with antibodies against mast cell and/or basophil antigens

For the detection of human mast cells, antibodies against the mast cell specific serine proteinase tryptase may be used (Irani *et al.*, 1989). Antibodies against kit, the receptor for stem cell factor, are also available, although this surface antigen is not specific for mast cells. It is also possible to stain for the expression of high-affinity IgE receptors, which occurs on both basophils and mast cells. Giemsa staining may be used to distinguish mast cells from basophils on a morphological basis.

Materials and equipment

anti-human tryptase (MAB 1222, Chemicon, Temecula, CA)
anti-kit/SCFR/CD117 (mouse or human) (Pharmingen, San Diego, CA)
anti-human high affinity IgER (MAB 103, Chemicon)
TBS (Tris buffered saline)
DAKO PAP kit system (DAKO, Glostrup, Denmark)
Immuno-mount (Shandon, Pittsburg, PA) and Permount (Fisher Scientific, Fair Lawn, NJ)

Method

1 Cytocentrifuged slides are fixed in acetone for 10 min.
2 Rinse in TBS 3 times.
3 Drip 2 drops of 3% H_2O_2 (bottle 1 included in the DAKO PAP kit system) on each slide and incubate for 5 min.
4 Rinse in TBS 3 times.
5 Block with normal rabbit serum diluted 1:1 in TBS and incubate for 15 min at room temperature.
6 Gently remove the blocking solution.
7 Incubate with primary antibody for 45 min at room temperature (include one slide without antibody as a negative control).
8 Rinse in TBS 3 times.
9 Drip 2 drops of secondary antibody (bottle 4) and incubate for 20 min.
10 Rinse in TBS 3 times.
11 Drip 2 drops of PAP (bottle 5) and incubate for 20 min.
12 Rinse in TBS 3 times.
13 Mix 3-amino-9-ethyl carbazole (AEC) solution: 1 drop of AEC (bottle 6) to 2 ml of buffer (bottle 7). For 5–8 slides 2 ml is enough. Add H_2O_2 (bottle 8). Mix. Add 50 µl to each slide and incubate for 10 min.

14 Rinse in TBS and then in tap water for 5 min.

15 Counterstain with Mayer's haematoxylin for 1.5 min.

16 Rinse and leave the slides in running tap water approximately 20 min.

17 Mount the slides.

Metachromatic staining

Both mast cells and basophils contain multiple prominent granules that stain from red to purple with blue aniline dyes, thus exhibiting metachromasia. Metachromasia is now attributed to highly sulphated proteoglycans within mast cell and basophil granules. Toluidine blue at acid pH is used for the visualisation of both basophils and mast cells.

Materials:

Toluidine blue (Sigma)

12 M HCl

Permount

Method

1 Make a 0.5% toluidine blue solution in 0.5 M HCl.

2 Apply the solution onto a cytospin preparation.

3 Incubate 60 min.

4 Rinse gently with tap water.

5 Mount a coverslip on the slide with Permount.

Enzymatic staining

The sensitive and specific tryptase substrate, Z-Gly-Pro-Arg-4-methoxy-2-naphthylamide (MNA), applied to an enzyme-histochemical staining technique with either Fast Garnet GBC or Fast Black K as the chromogen has produced an easy and rapid staining method for mast cells (Harvima et al., 1988; Kaminska et al., 1996). This method allows detection of mast cells in less than 20 minutes.

Materials:

Z-Gly-Pro-Arg-MNA (Bachem, Bubendorf, Switzerland), 20 mM stock solution dissolved in dimethylformamide (stable at −20°C)

Fast Garnet GBC or Fast Black K salt (Sigma) at 4 mg/ml in double distilled
water (ddH$_2$O) (store in aliquots at −20°C)
0.5 M Tris-HCl, pH 7.5

Method

1 Cytospin preparations are fixed in acetone for 10 min.
2 Mix 20 µl Z-Gly-Pro-Arg-MNA, 40 µl 0.5 M Tris-HCl, pH 7.5, 25 µl
  Fast Garnet GBC or Fast Black K, and 120 µl ddH$_2$O immediately before
  use.
3 Centrifuge the substrate solution briefly to clear.
4 Aliquots (c. 200 µl) of the substrate solution are applied onto cytospin
  preparations.
5 Follow the enzymatic reaction under a microscope for 5–15 min. Wash
  the slides with tap water. Count positive cells. The red dye of Fast Garnet
  GBC is labile and can be washed away with an overnight incubation in
  15% Tween-20. The dark blue dye of Fast Black K can be stabilised by
  incubating the cytospins in 1–2% CuSO$_4$ for 5–10 min.

Summary

Techniques are detailed for culturing murine and human basophils and mast
cells. Basic protocols are also described for the detection of cultured mast
cells and basophils. These methods will hopefully aid those wishing to
examine mast cells and basophils.

Applications

The development of techniques to culture mast cells and basophils has had a
significant impact on the study on the biology of these cells. There are two
main applications for these *in vitro* cell cultures:

1 Studies on mast cell and basophil differentiation. Using the protocols
  given in this chapter, studies may be performed on cell commitment,
  genes involved in mast cell and/or basophil differentiation, the regulatory
  effect of cytokines and growth factors in the differentiation process, etc.
2 *In vitro* developed mast cells and basophils, used for further studies on their
  cell biology. The difficulties in obtaining large numbers of pure tissue mast
  cells or peripheral blood basophils have made their *in vitro* developed
  counterparts important cell sources for further studies on their biology.

Thus, studies on activation and exocytosis, adhesion, cytokine expression, gene cloning, migration, receptor expression and intracellular signal transduction pathways, and survival can be furthered with these cells.

Troubleshooting

Mouse mast cells

Development of bone-marrow-derived mouse mast cells should be reproducible. If mast cells cannot be detected within 2 weeks of culture there may be problems with the fetal calf serum, or WEHI conditioned medium, or contamination.

Human basophils and mast cells

There may well be differences in basophil/mast cell development in cultures of cells from different donors. The outcome is more reproducible if CD34$^+$ cells are used, thereby reducing the number of accessory cells in the cultures that may secrete factors having a negative effect on basophil/mast cell differentiation. If cultures fail to develop, suspect contamination by microorganisms, or that the growth factors used are not biologically active.

Acknowledgements

We wish to thank Dr Arnold Kirshenbaum, LAD, NIAID, NIH, Bethesda, MD, for providing the photomicrographs shown in Fig. 8.1. G.N. is supported by The Swedish Cancer Society.

References

Eklund, K. K., Ghildyal, N., Austen, K. F. & Stevens, R. L. (1993). Induction by IL-9 and suppression by IL-3 and IL-4 of the levels of chromosome-14-derived transcripts that encode late-expressed mouse mast cell proteases. *J. Immunol.*, **151**, 4266–73.

Galli, S. J., Zsebo, K. M. & Geissler, E. N. (1994). The kit ligand, stem cell factor. *Adv. Immunol.*, **55**, 1–96.

Gurish, M., Ghildyal, N., McNeil, H. P., Austen, K. F., Gillis, S. & Stevens, R. L. (1992). Differential expression of secretory granule proteases in mouse mast cells exposed to interleukin-3 and c-kit ligand. *J. Exp. Med.*, **175**, 1003–12.

Harvima, I. T., Naukkarinen, A., Harvima, R. J. & Horsmanheimo, M. (1988). Enzyme- and immunohistochemical localization of mast cell tryptase in psoriatic skin. *Arch. Dermatol. Res.*, **281**, 387–91.

Irani, A. A., Bradford, T. R., Kepley, C. L., Schechter, N. M. & Schwartz, L. B. (1989). Detection of MC_T and MC_{TC} types of human mast cells by immuno-histochemistry using new monoclonal anti-tryptase and anti-chymase anti-bodies. *J. Histochem. Cytochem.*, **37**, 1509–15.

Irani, A. A., Nilsson, G., Miettinen, U., Craig, S. S., Ashman, L. K., Ishizaka, T., Zsebo, K. M. & Schwartz, L. B. (1992). Recombinant human stem cell factor stimulates differentiation of mast cells from dispersed human fetal liver cells. *Blood*, **80**, 3009–21.

Kaminska, R., Harvima, I. T., Naukkarinen, A., Nilsson, G. & Horsmanheimo, M. (1996). Alterations in mast cell proteinases and protease inhibitors in the progress of cutaneous herpes zoster infection. *J. Pathol.*, **180**, 434–40.

Kirshenbaum, A. S., Goff, J. P., Dreskin, S. C., Irani, A. M., Schwartz, L. B. & Metcalfe, D. D. (1989). IL-3-dependent growth of basophil-like cells and mast-like cells from human bone marrow. *J. Immunol.*, **142**, 2424–9.

Kirschenbaum, A. S., Goff, J. P., Kessler, S. W., Mican, J. M., Zsebo, K. M. & Metcalfe, D. D. (1992). Effect of IL-3 and stem cell factor on the appearance of human basophils and mast cells from $CD34^+$ pluripotent progenitor cells. *J. Immunol.*, **148**, 772–7.

Mitsui, H., Furitsu, T., Dvorak, A. M., Irani, A. A., Schwartz, L. B. Inagaki, N., Takei, M., Ishisaka, K., Zsebo, K. M., Gillis, S. & Ishizaka, T. (1993). Development of human mast cells from umbilical cord blood cells by recombi-nant human and murine c-kit ligand. *Proc. Natl. Acad. Sci. USA*, **90**, 735–9.

Nilsson, G., Blom, T., Harvima, I., Kusche-Gullberg, M., Nilsson, K. & Hellman, L. (1996). Stem cell factor-dependent human cord blood derived mast cells express α- and β-tryptase, heparin and chondroitin sulphate. *Immunology*, **88**, 308–14.

Rottem, M., Goff, J. P., Albert, J. P. & Metcalfe, D. D. (1993). The effects of stem cell factor on the ultrastructure of Fc epsilon RI+ cells developing in IL-3-dependent murine bone marrow-derived cell cultures. *J. Immunol.*, **151**, 4950–63.

Saito, H., Hatake, K., Dvorak, A. M., Leiferman, K. M., Donnenberg, A. D., Arai, N., Ishizaka, K. & Ishizaka, T. (1988). Selective differentiation and proliferation of hematopoietic cells induced by recombinant human interleukins. *Proc. Natl. Acad. Sci. USA*, **85**, 2288–92.

Saito, H., Ebisawa, M., Tachimoto, H., Shichijo, M., Fukagawa, K., Matsumoto, K., Iikura, Y., Awaji, T., Tsujimoto, G., Yanagida, M., Uzumaki, H., Takahashi, G., Tsuji, K. & Nakahata, T. (1996). Selective growth of human mast cells induced by steel factor, IL-6, and prostaglandin E_2 from cord blood mono-nuclear cells. *J. Immunol.*, **157**, 343–50.

Sillaber, C., Sperr, W. R., Agis, H., Spanblöchl, E., Lechner, K. & Valent, P. (1994). Inhibition of stem cell factor dependent formation of human mast cells by inter-leukin-3 and interleukin-4. *Int. Arch. Allergy Immunol.*, **105**, 264–8.

Valent, P., Schmidt, G., Besemer, J., Mayer, P., Zenke, G., Liehl, E., Hinterberger,

W., Lechner, K., Maurer, D. & Bettelheim, P. (1989). Interleukin-3 is a differentiation factor for human basophils. *Blood*, **73**, 1763–9.

Valent, P., Spanblöchl, E., Sperr, W. R., Sillaber, C., Zsebo, K. M., Agis, H., Strobl, H., Geissler, K., Bettelheim, P. & Lechner, K. (1992). Induction of differentiation of human mast cells from bone marrow and peripheral blood mononuclear cells by recombinant human stem cell factor/kit-ligand in long-term culture. *Blood*, **80**, 2237–45.

Index

180

enzyme digestion 16
homogenisation 16
embryonic stem (ES) cells 6–7
  antigen expression 15
  isolation and culture
    cell culture 8–9
    cell harvesting 9
    formation of embryoid bodies *see*
      embryoid bodies
    generation of undifferentiated
      aggregates 9–12
      hanging drop method 10–11
      high density culture 11–12
    see also haematopoietic differentiation
Epstein–Barr virus *see* EBV-transformed
  B cells
erythroid cells 15
erythropoiesis 15
ES cells *see* embryonic stem cells

F4/80 122
FACS analysis *see* fluorescence activated
  cell sorter analysis
Fas 26
fascin (p55) 27
'feeder cells' 73–4, 75, 76, 82, 162
fetal calf serum 97, 99, 127
fibroblasts 60
fibronectin 46
flow cytometric analysis
  of dendritic cells 27, 45
  of haematopoietic precursors and
    progeny 15–16
  of NK cells 153–4
  of thymocytes 67–9
Flt-3 ligand 25
fluorescein isothiocyanate (FITC) 67,
  83, 107, 109, 153
fluorescence activated cell sorter (FACS)
  analysis
  B lymphocytes 106–12
    antibodies for murine/human B cells
      107–8
    cell cycle progression and apoptosis
      110–12
    sorting 109–10
    staining 108–9

of fluorochrome-labelled dendritic cells
  42
monocytes and macrophages 129–30,
  131, 138, 139
NK cell cloning 162
T cells 83
thymocytes 69
fluorochrome-labelled dendritic cell
  migration 37–8, 40–2
formyl peptides 43

gamma camera 39–40
giant cell fusion assay 130–3
Giemsa staining 174
Z-Gly-Pro-Arg-methoxy-2-
  naphthylamide (MNA) 175–6
GM1 150
granulocyte–macrophage colony
  stimulating factor (GM-CSF)
  25, 29, 30, 31, 33, 34, 35–6,
  37, 42, 45
granulocyte–monocyte stem cells 120

haematopoietic differentiation 5–22
  analysis of differentiated progeny
    analysis of colony forming potential
      17–20
    flow cytometric analysis 15–16
    gene expression studies 17
    histology 14–15
    reconstitution of recipient animals
      20
  embryoid bodies 12–14
  embryonic stem cell isolation and
    culture 8–12
  haematopoietic progenitors
    colony forming potential 17–20
    re-constitution of recipient animals
      with 20
  methodologies and reagents 7–8
  stem cells and types of 5–7
haematopoietic stem cells (HSC) 5–7,
  15
heat stable antigen (HSA) 94
HIV-1 43
HLA polymorphism 71
HLA-DR 130, 131

Printed in the United States
By Bookmasters